# NORTH by NUUK

GREENLAND AFTER ROCKWELL KENT

# NORTH by NUUK

## GREENLAND AFTER ROCKWELL KENT

DENIS DEFIBAUGH

This book is based upon work supported by the National Science Foundation under Award No. PLR-1524176. Any opinions, findings, and conclusions or recommendations expressed in this book are those of the authors and do not necessarily reflect the views of the National Science Foundation.



Published and distributed by:
RIT Press
90 Lomb Memorial Drive
Rochester, New York 14623-5604
http://ritpress.rit.edu

Printed in the U.S.A.

ISBN 978-1-939125-72-9

Library of Congress Cataloging-in-Publication Data

Names: Defibaugh, Denis, author, photographer. | Defibaugh, Denis. Greenland after Rockwell Kent. Kalâtdlisut.
Title: North by Nuuk : Greenland after Rockwell Kent / [author] Denis Defibaugh.
Description: First edition. | Rochester, New York : RIT Press, 2019. | Includes an essay translated into Kalâtdlisut. | English and Kalâtdlisut. | Summary: "North by Nuuk is an intimate, contemporary look at the people and the social and primal geographic landscapes of Greenland. Photographer Denis Defibaugh presents his journey from Nuuk to the settlement of Illorsuit, 300 miles north of the Arctic Circle, following Rockwell Kent's earlier footsteps and offering a fresh look at timeless Greenland"-- Provided by publisher.
Identifiers: LCCN 2019040074 | ISBN 9781939125729 (hardcover)
Subjects: LCSH: Greenland--Pictorial works. | Greenland--Social life and customs. | Kent, Rockwell, 1882-1971--Travel--Greenland. | Defibaugh, Denis--Travel--Greenland. | Travel photography--Greenland. | Repeat photography--Greenland.
Classification: LCC G743 .N7185 2019 | DDC 998/.2--dc23
LC record available at https://lccn.loc.gov/2019040074

# Contents



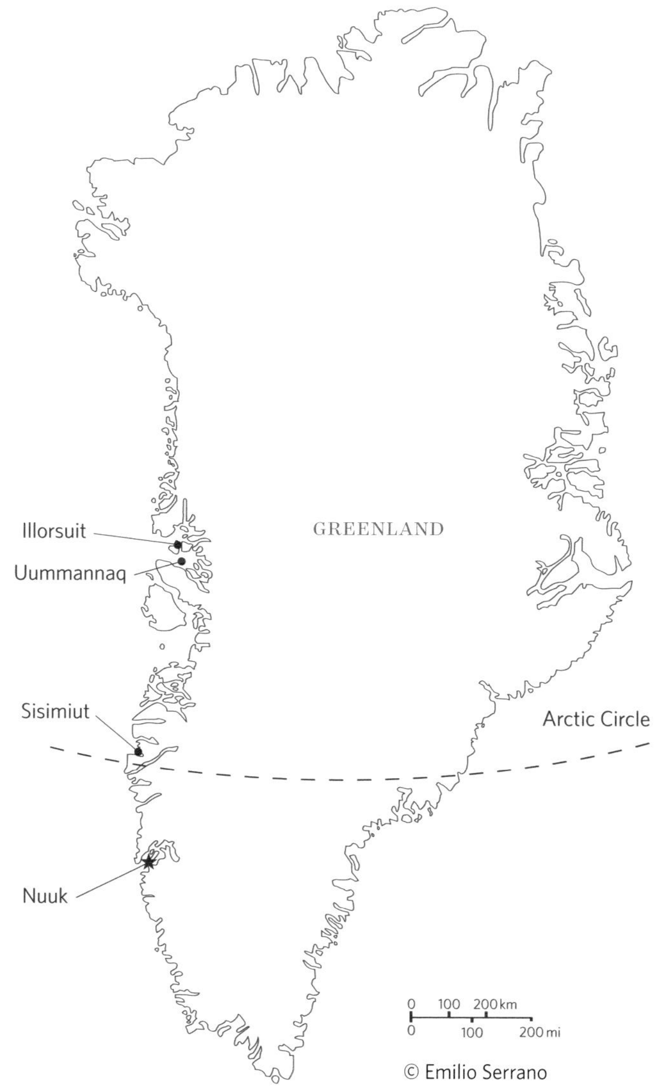

Illorsuit
Uummannaq
GREENLAND
Sisimiut
Arctic Circle
Nuuk
0 100 200 km
0 100 200 mi
© Emilio Serrano

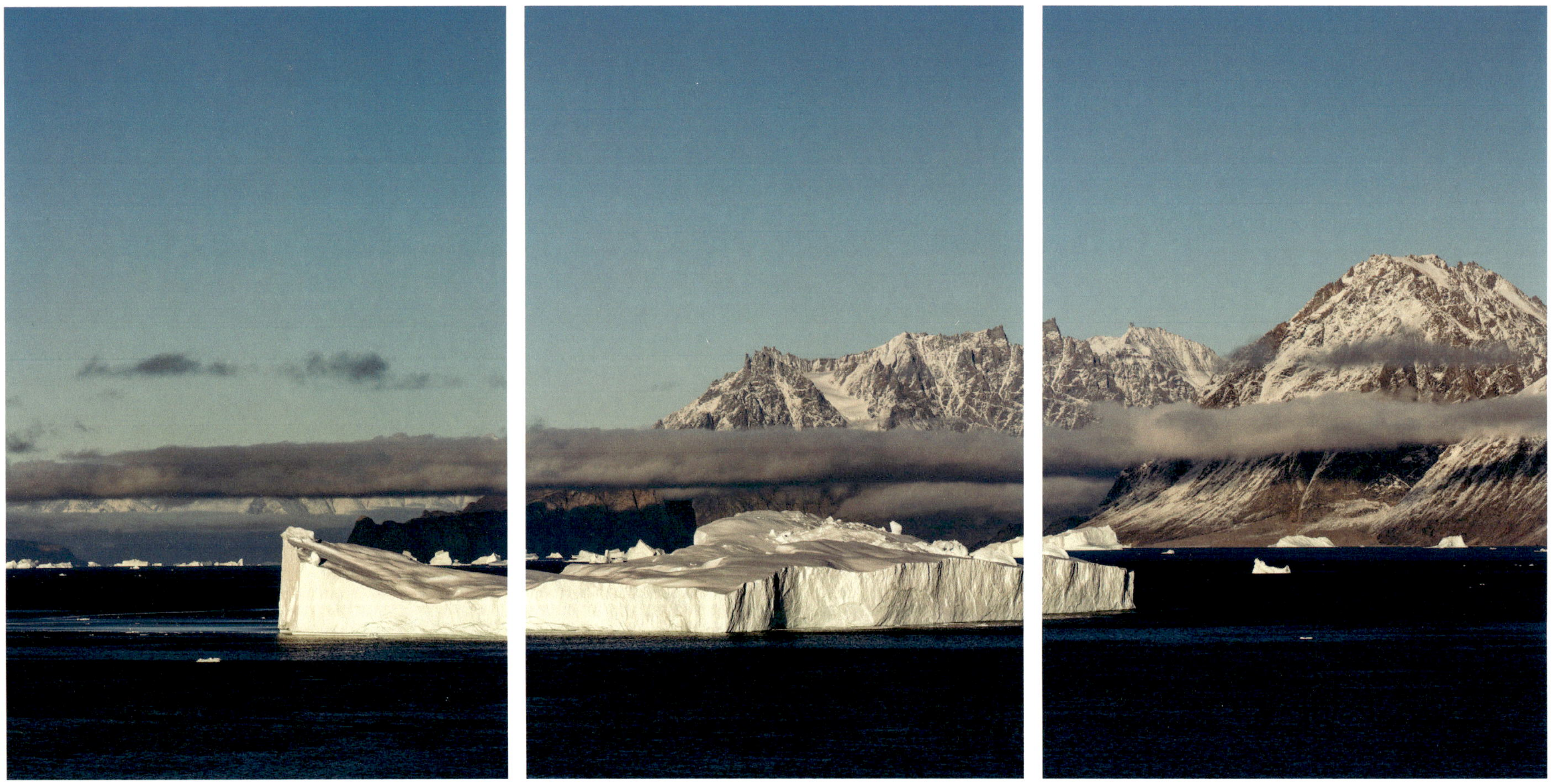

ILLORSUIT 2016
An iceberg dominates the Illorsuit Sound as a band of dark clouds stripes the Sermeq Qiterleq Mountain.

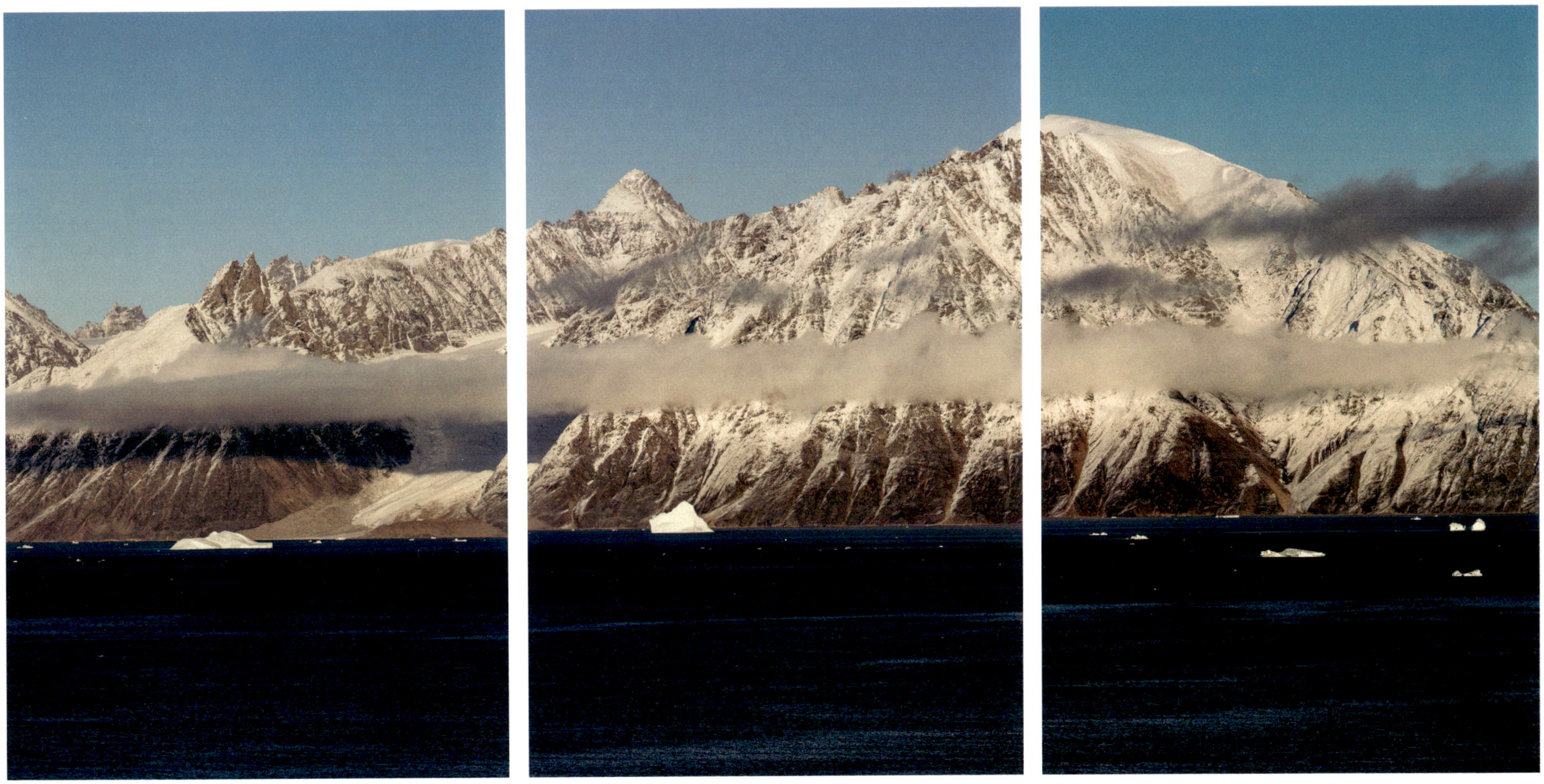

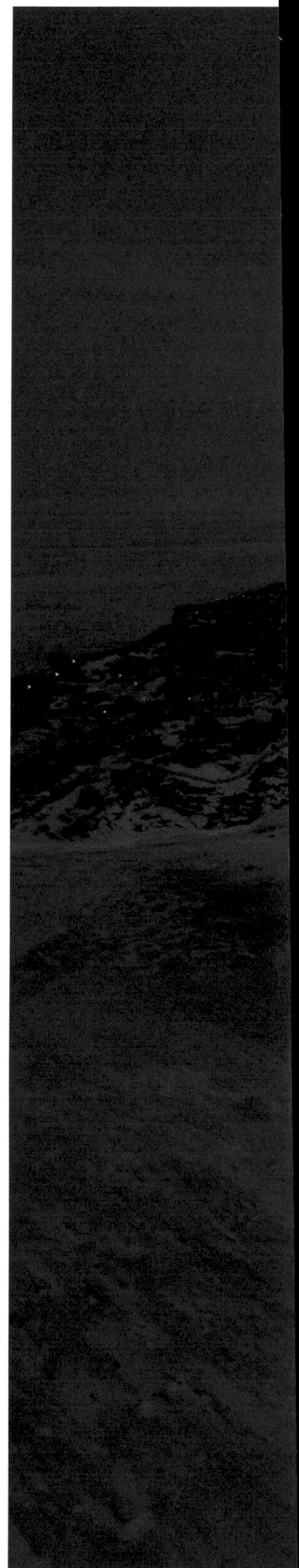

UUMMANNAQ 2017
Uummannaq Island, Uummannaq Mountain, and the frozen Uummannaq Fjord.

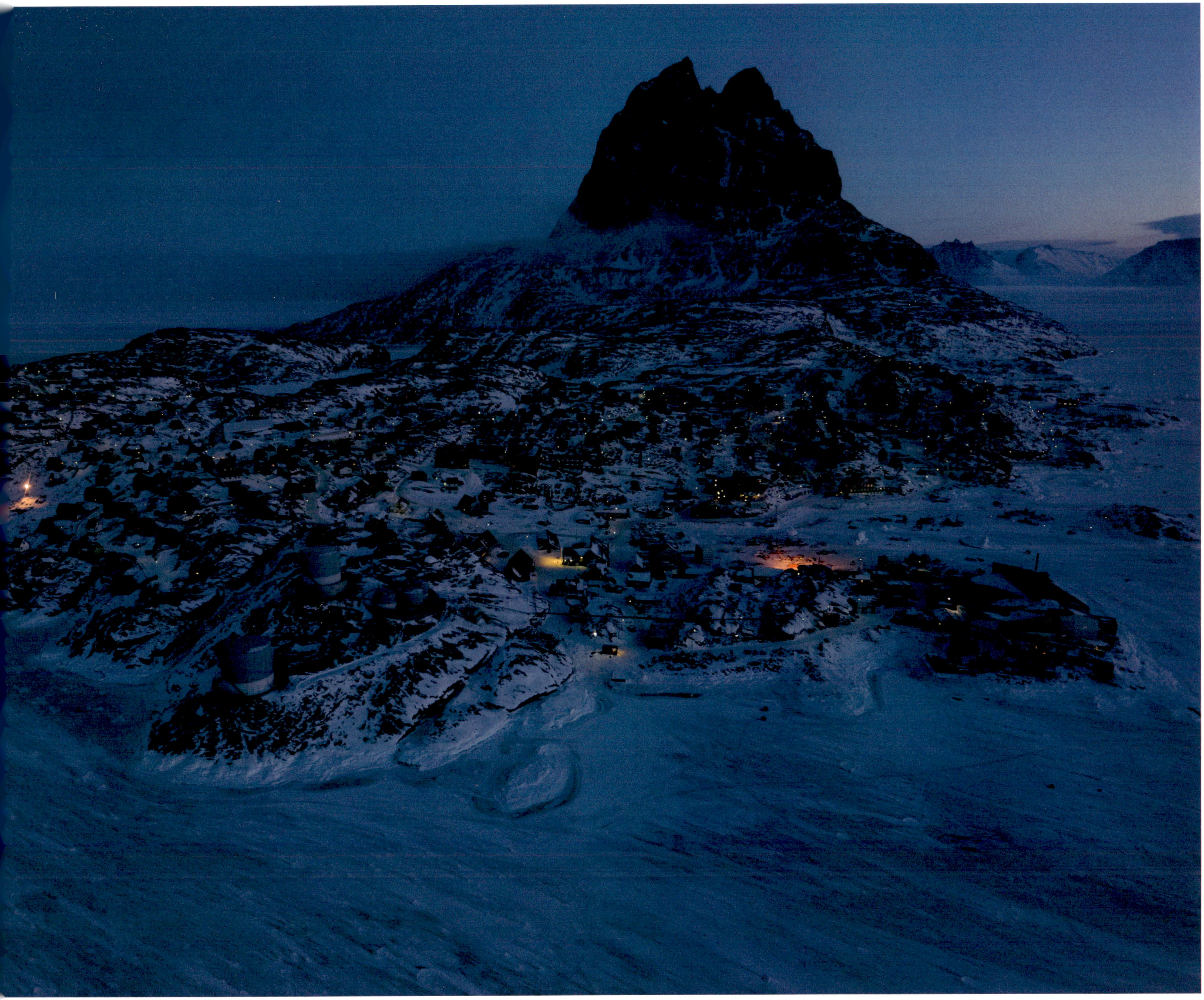

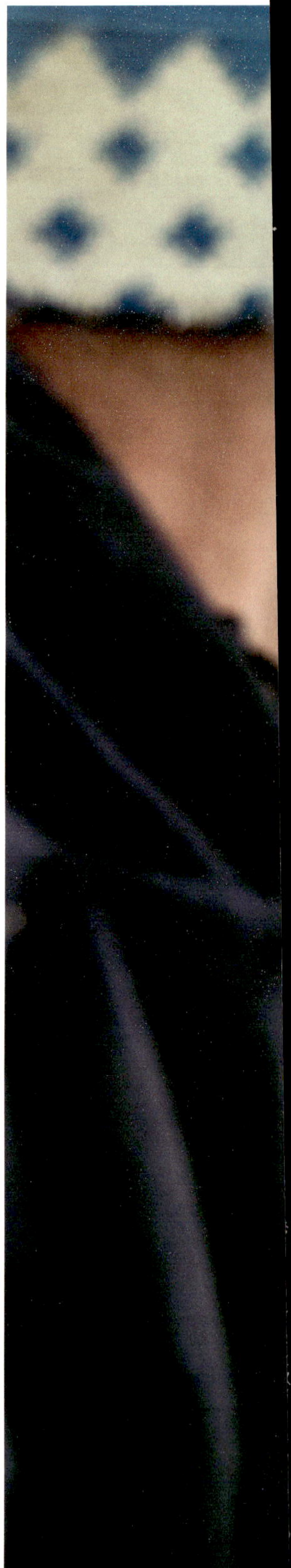

UUMMANNAQ FJORD 2017
Hunters Pujunnguaq Hammond and Karl Nielsen dog sledging to their fishing holes. Temperature is about -10 degrees F.

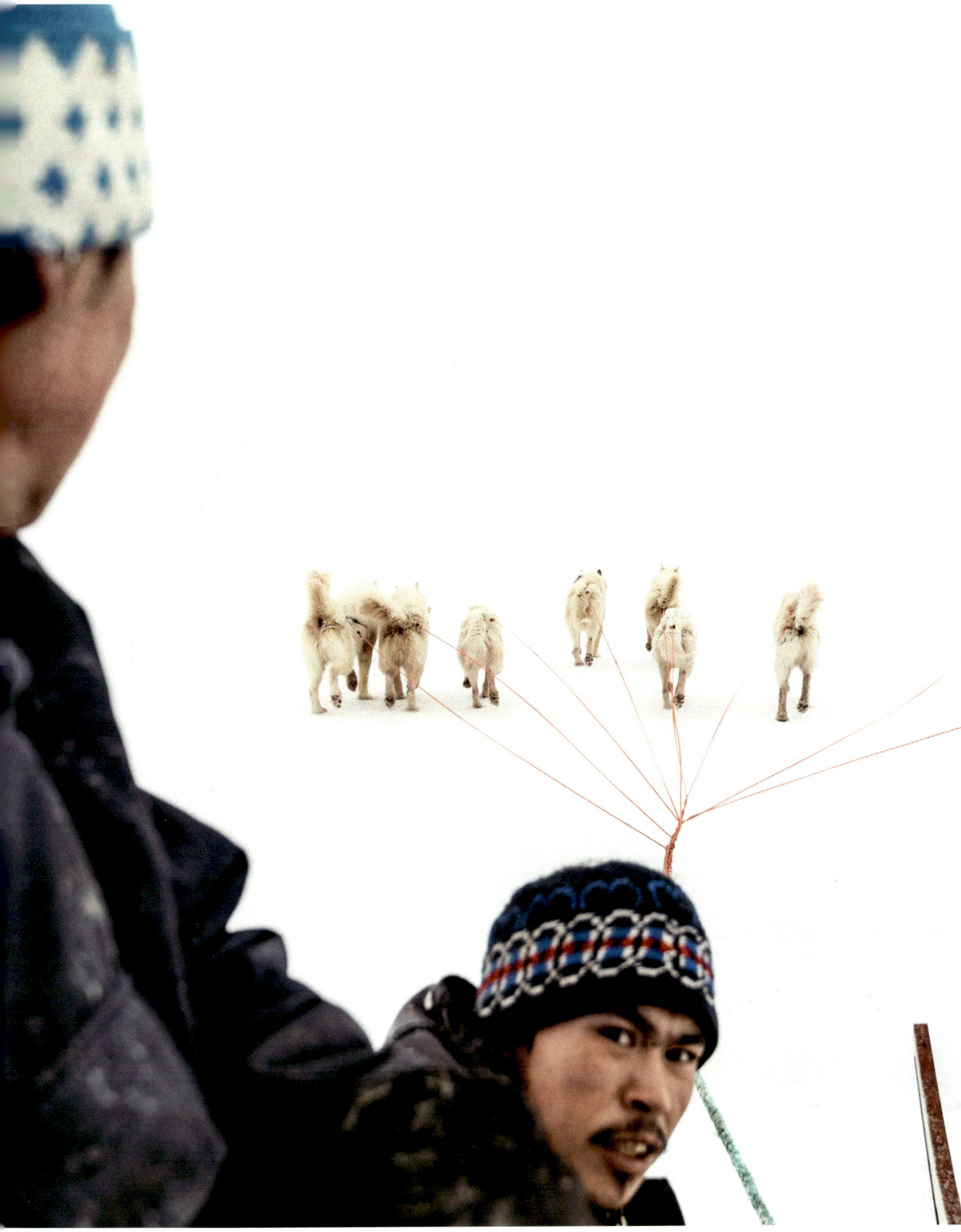

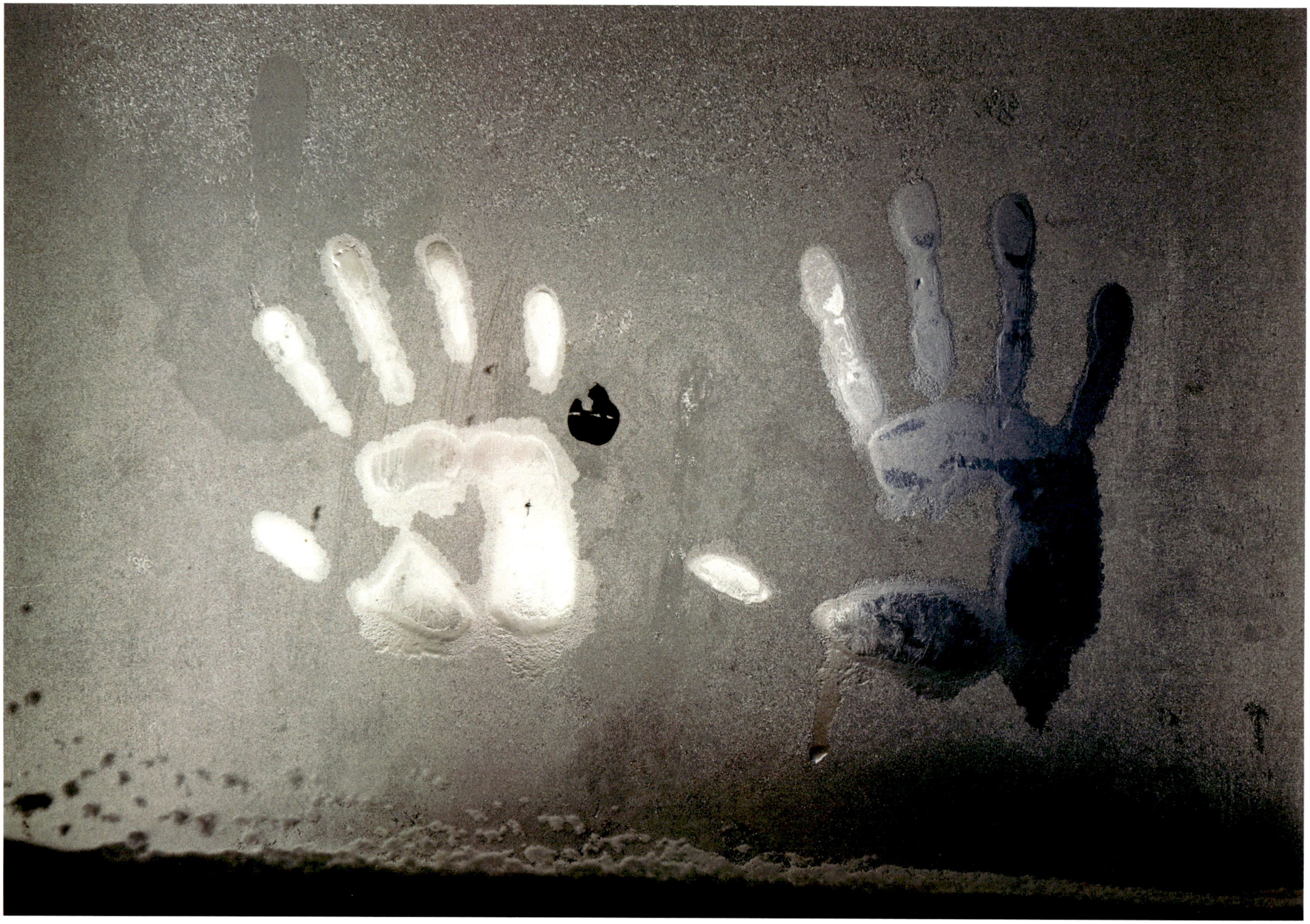

ILLORSUIT 2017
Embedded on an icy window pane are hand imprints of two children.

# Foreword

**By the time Rockwell Kent set off for Greenland in 1929 on his 33-foot sailboat, he already had felt the pull of landscapes that were cold and remote. He had sailed the Beagle Channel off Tierra del Fuego and, with his son, had spent a harsh winter on Alaska's Fox Island.**

Inspired by early readings of Viking sagas and his own distaste for confinement, his talent and ambition fueled his peregrinations. He didn't "saunter," as Thoreau might have recommended, but rather made mad dives into the unknown. "I crave snow-topped mountains, dreary wastes, and the cruel North Sea with its far horizons at the edge of the world," he wrote in his Alaska journal, *Wilderness*.

The craving became almost an addiction to the far north; a possession I also have felt. Once you go above the Arctic Circle in Greenland, you keep going back for more: sea ice and the horizon, sledge dogs and subsistence villages—that's the Siren who calls and calls. Bad luck turned into luck for Kent, when his sailboat dragged anchor in a fjord south of Nuuk and broke up on rocks. While waiting for it to be repaired, he made his first forays across ice-scoured rocks of the many-fingered fjords and quickly understood that nothing could stop him from returning.

Return he did in 1931, after another piece of good luck: He'd met the two most famous Greenland explorers, Knud Rasmussen and Peter Freuchen, on a ship to Copenhagen and they helped Kent organize a stay in an isolated and exceptionally beautiful village on *Ubekendt Eyland*—Unknown Island—50 miles northwest of the bustling village of Uummannaq. Set back from the Davis Strait in a matrix of fjords, Uummannaq is a water-bound fortress dominated by a heart-shaped mountain almost 4,000 feet high, with houses bolted to rock sloping down to the sea. It was there that Kent gathered supplies and hired a local woman named Salamina as his *kifak*—his housekeeper and later his lover—and made his way by boat to Illorsuit, Unknown Island's only village.

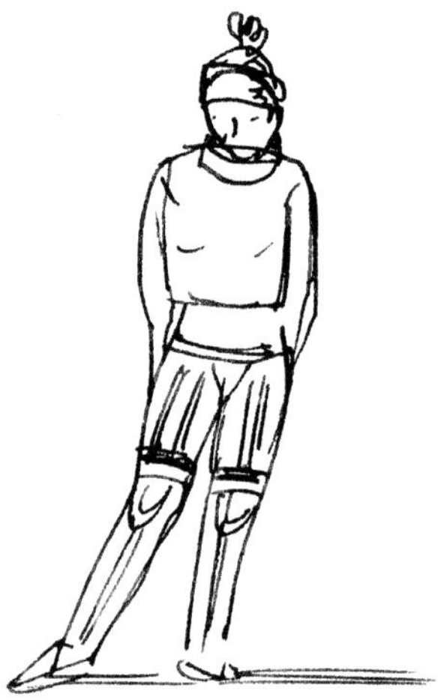

Sometimes a place is so generous in its offerings that it's hard to know how to give back. Kent painted avidly and participated in village life. He built a house; brought music; oversaw the construction of a dance house; made deep friendships; and in every season, whether by dog sledge or skiff, went north up the Illorsuit Sound to other remote islands to paint, draw, and photograph. His time there, amounting to more than two years, became his physical and spiritual canvas, his cold paradise. Of that time, he wrote in his book, *Greenland Journal*: "It was perhaps the happiest and certainly most productive of my life."

Eighty-six years later, the photographer Denis Defibaugh discovered a collection of glass plate photographs made by Kent and was determined to follow Kent's travels in Greenland. In winter and summer, Defibaugh photographed in Nuuk, Uummannaq, Illorsuit, and Sisimiut, and in the resulting book, *North by Nuuk*, we see Defibaugh's contemporary Greenland alongside Kent's views of life in the 1930s. Much has changed; much is the same.

Kent spent most of his hours painting and drawing. His photographs were casual snapshots of daily life made with a 35mm Leica, and from those negatives, he created 3 x 2 positive glass negatives hand-colored with dyes; Defibaugh's digital images are elegant. Landscapes and portraits—both the old and the new—can be viewed side by side. Kent's close-up portrait of Salamina, wearing traditional clothing, is luminous, with an austere sensuality that matches the place itself. Defibaugh's portrait of a seated Uummannaq woman smoking is coquettish, ironic, and modern in both the furnishings and her clothing; even the painting on the wall behind her head appears to be of sperm wiggling through rounds of bright colors.

A quarter of a century ago when I arrived in Uummannaq—at the behest of two travelers I met while weathered in at Baffin Island's airport—modernity was less conspicuous. I spent part of a winter in a cold house and the all-dark days and nights reset my circadian rhythms. The *pulaar*—nightly visits to Ann and Ole Jorgen's house for later dinners or coffee and cake at midnight—were the godsend.

Uummannaq, like all northern Greenland villages, had only two cars and one narrow road. The houses, whose windowsills were lined with candles, were accessed only by steep walking paths. There were 6,000 sledge dogs and 1,400 people, and the nightly chorus of howling sang us to sleep. With no television, no internet, and no cell phones, traditional stories were still told, the same ones that had been passed down for thousands of years, all part of a single culture and language (with vast dialect differences) that spanned the polar north—so the stories told in Uummannaq or Illorsuit were the same ones told in Point Hope, Alaska or Vankarem, Chukotka. But as everywhere, with the coming of television, storytelling took a nosedive—except, my hunter friends told me, when they were out for weeks or months on the sea ice or on kayaks in the fjords hunting for food.

When I look at Kent's paintings and photographs made in Illorsuit, I'm reminded viscerally of my time there. In 1996, I had hitched a ride on a fishing boat after writing to and receiving an invitation to visit from Hans, the only English-speaking resident of the village and a Danish dropout who had gone there one summer and stayed. I wanted to live where Rockwell Kent had painted and written his wondrously lyrical books, *Salamina* and *Greenland Journal*, and hoped to rent his house for the summer, only to find out at 2 in the morning when we finally arrived that the house had been taken down.

Illorsuit was a handful of 25 houses set on a horseshoe-shaped beach that faces north across a six-mile-wide strait to the uninhabited island of Upernavik beyond. Icebergs big as mountains floated by. Village life delighted me as it had Kent. In *Salamina*, he wrote: "Igdlorssuit—July—and not a soul doing anything. Everyone was out of doors. It was a breathless sunlit afternoon." That stillness and bold beauty showed in his paintings. His photographs revealed the complete ease with which he slipped into Greenlandic culture.

Village life was not much different in the early 1990s than it had been in the 1930s. Families still slept together in large communal beds and kept irregular hours. Children roamed freely—there were no cars, no roads, nothing that could hurt them. As Kent had noticed, Inuit societies reveled in idleness until the moment they had to go hunting. Then they were all speed and precision.

On warm days Hans, his six-year-old daughter, Marie Louisa, and I went out in his skiff to gather tern eggs and camped and fished on Karrat Island, where Kent loved to paint. I ate raw liver—still steaming—on the beach where a seal had been butchered. Near the village, I wandered the hills with a gang of children following me. I was surprised to find that no one bathed: there were no facilities in which to get clean. Sometimes Hans hauled water in from a spring and heated it on the stove so I could wash my hair. When I found a thin waterfall on one of the walks with Marie Louisa, we took off our clothes and let the frigid water spill down our bodies. She had never seen a shower, much less a car, a horse, a tree, a two-story building, or city lights, nor had she ever ridden on an airplane. Her modes of transportation were a mix of new and old: short Norwegian skis and her own dog sledge pulled by three dogs.

The paintings Kent made in the summer months in Illorsuit were bold, with a palette, as I later wrote, "made of blended mist and ice, aquamarine and gray and the watery blue-veined whiteness of a world without night." He had a keen eye for Arctic light, and his architectural studies showed in the way triangular shadows cartwheeled across barren landscapes. His photographs, however, were homey and affectionate: group portraits of his new friends and their traditional houses made of turf and stone with whale-rib roof beams and seal-gut windows. Those houses were gone by the time I arrived, replaced by Danish-style wooden houses for only one reason: to eradicate tuberculosis, which had been rampant.

One of Kent's photos shows hunters at camp with the dog sledges and a skin boat called an *umiak* in the background. Skin boats were large and capacious, rowed only by women, and were used to hold the catch—whatever it was. Those boats were gone when I arrived, but I was able to interview Ann Hansen in Illorsuit, the last of the eight women who rowed the village umiak during reindeer-hunting season. "We went all the way down to Kangerlussuaq to get the reindeer," she told me. "It took several days to get there. We carried the meat home. Other times we rowed to those islands north of Illorsuit to collect eggs and hunt birds. Rowing was hard day after day. We were very strong."

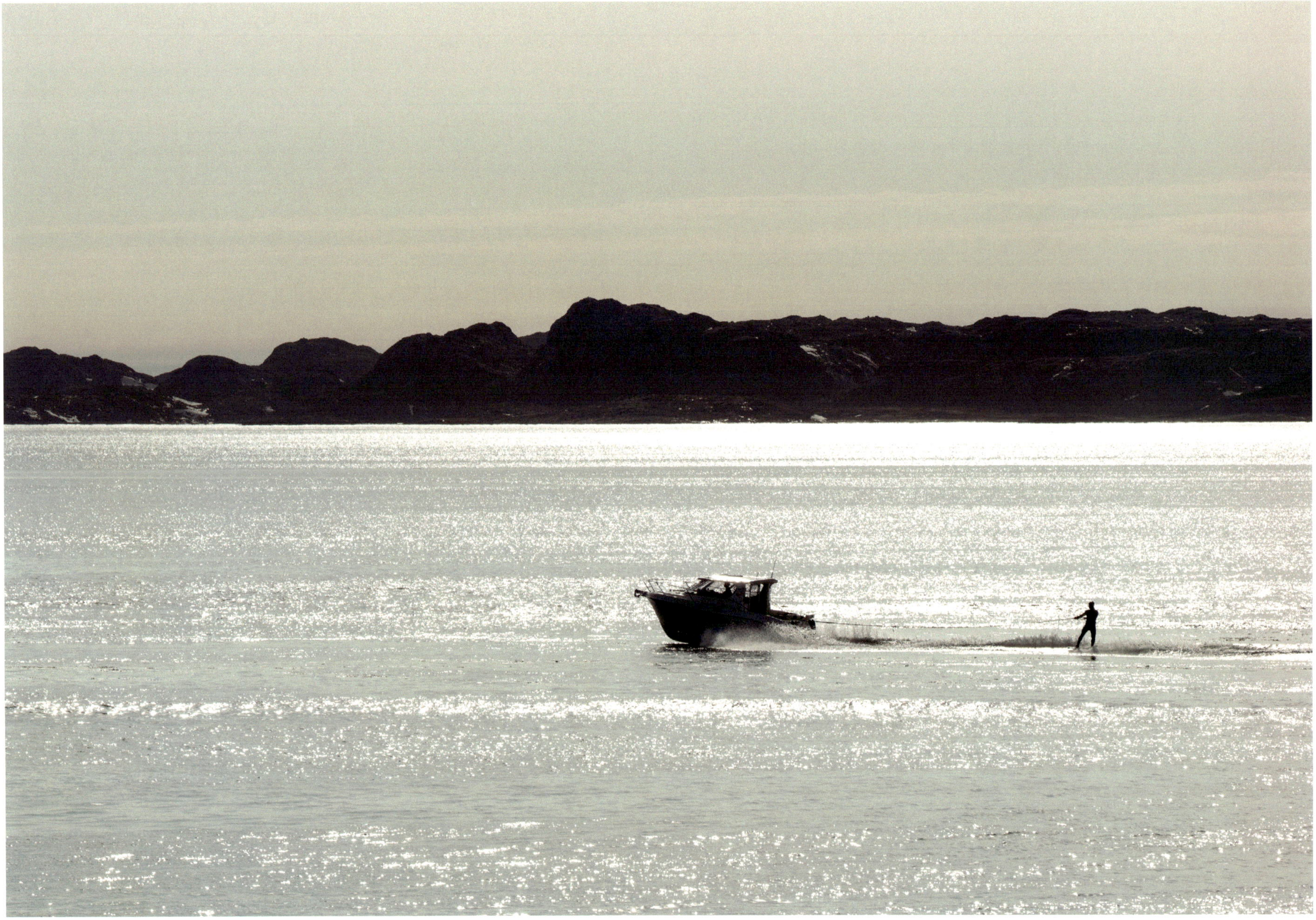

NUUK 2016

Water skier in one of Nuuk's fjords. The water is barely above freezing. This is a rare sight. The only time I ever saw a water skier in Greenland.

My fondness for the village redoubled when Aliberti, a hunter I'd traveled with from Ikerasak, found himself adrift on a piece of ice and ended up 50 miles from home in front of Illorsuit. It was Christmas Eve; he had been drifting for a month. Someone came outside and saw him, and the good people of the village paddled out and saved his life. His hair had fallen out from hunger and stress. His parents already had held his in-absentia funeral, and when the villagers took him back to Ikerasak, his parents thought he was a ghost and ran away. Hans's neighbor said, "We take care of each other, live together, hunt together, have fun together, otherwise we'd all be dead."

Nuuk is the capital city of Greenland, well below the Arctic Circle. Even when Rockwell Kent crash-landed, it was a town of some modernity with obvious Scandinavian influence. Nuuk has a university, a museum, a cooking school, a school for helicopter mechanics, and a hospital, and is the center of home-rule government. One of the first women I met in Uummannaq, and with whom I traveled by dog sledge in the far north, was Aleqa Hammond, who later became Greenland's first female prime minister.

Its rows of apartment buildings have grown since I was there in 1993, and like any city, its sophistication can belie the usual problems of drugs, joblessness, and alcoholism. Defibaugh's photograph shows youngsters playing handball against a wall, and the portrait of a young woman wearing a black blazer, black leggings, tennis shoes, a white shirt, and a magenta/pink bowtie and shades is hip and cool. In Kent's photograph of Sisimiut, another sizeable town farther north, we see a sturdy two-story red church overlooking the harbor and the ship, *Disko*, which brought Frances Kent, Rockwell's wife, to Greenland. In contrast is Defibaugh's image of two girls wearing traditional costumes in church checking out their iPhones.

Contemporary images of women elsewhere dressed in seal skin suits or North Face winter gear are more common above the Arctic Circle, and children abound, always. Many of the landscapes of both Kent and Defibaugh are alike yet different. In a wintry Uummannaaq, we see a young couple pushing a pram across the sea ice in front of town and kids playing ice soccer. In Kent's time, mothers might have been carrying babies in *amuts*, but ball games have always been around, although the balls might have been made of walrus skulls wrapped in skins.

Culture is always changing, the new merging with the old. What I loved most about Greenland's subsistence hunters in the north was the disciplined way they took of the modern what they needed but insisted on using dog sledges, wearing skins, and living much as they had in previous years. When I asked why, they said, "The old ways work better." The fact that dog sledges are now made of imported Danish wood (no trees grow in Greenland) instead of rattling bones and antlers, and dog harnesses are nylon instead of bearded seal skins (which the dogs kept eating) is of little importance, because the *language* has stayed intact: the words that tell you who you are, and how to live, and how to behave, and what existence means. They are still there. The change that has torn the culture apart is the loss of sea ice. "Everything we know has to do with ice," one hunter told me. Without it, lives change drastically, and a sense of cultural loss prevails. I was shocked when I looked down while flying north past Uummannaq and Illorsuit one cold February in 2007: there was no ice at all. The old Inuit sense of identity came from the word *Sila*, meaning the power of nature, every kind of weather, and the consciousness of all sentient beings as one and the same. As the climate changes, perhaps that understanding will be their guide.

*Gretel Ehrlich*
*May 28, 2019*

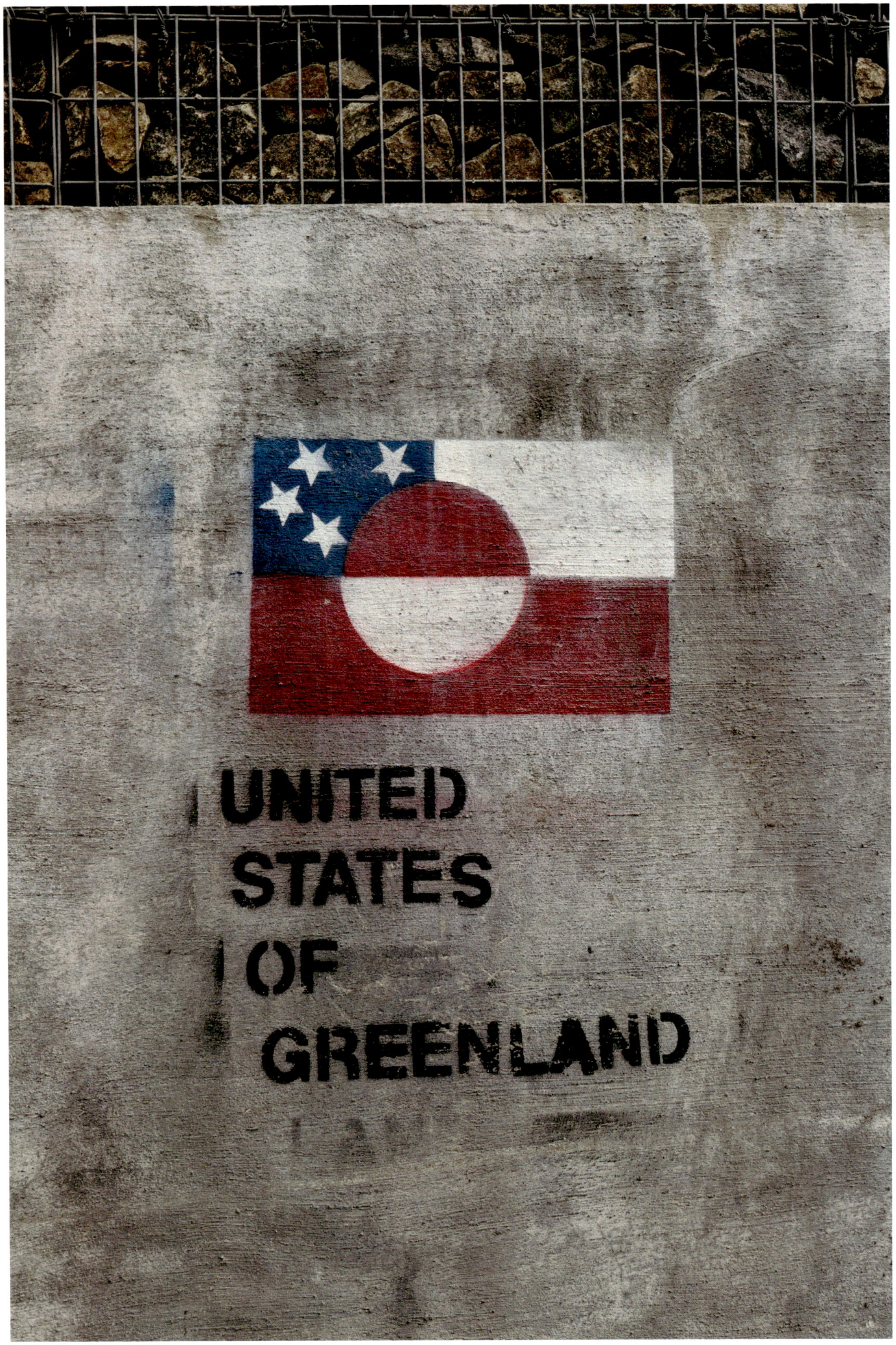


SISIMIUT 2017

Commentary on the relationship between Greenland and the United States as expressed by a graffiti artist. Historically, the two nations collaborated on scientific exploration and research and Greenland hosted US military bases.

# Greenland after Rockwell Kent: Illorsuit 1931 and 2016

DENIS DEFIBAUGH
*Greenlandic translation by Axel Jeremiassen*

*As the old hunter walked through the pearly gates and approached the Almighty, he said, "God, I only have one question for you: Why did you put mosquitoes in Greenland?" God looked at the elderly man with a smile that only He can make and said, "So Greenland doesn't compete with Heaven."*

**Illorsuit 1931 and 2016**

The anticipation of flying into Illorsuit for my first time in a classic Air Greenland Bell 212 helicopter was worth all of the five flights and three days from Stratton Air National Guard Base in New York to Kangerlussuaq to Ilulissat to Qaarsut to Uummannaq and now—finally—to Illorsuit. I was as excited as a kid in a candy store as we boarded the helicopter and jockeyed for window seats for the 30-minute flight. We lifted off, whipped around Uummannaq Island, and headed north as we entered the widest stretch of the Uummannaq Fjord. The stark white contrast of the massive sculpted icebergs against the deep cobalt blue of the fjord created dimension within the rolling seawater's surface. The mountainous snowcapped range of Greenland's mainland could be seen to the east, along with the seemingly infinite Greenland ice cap; and to the west, the southern tip of *Ubekendt Ejland*—Unknown Island, home to the settlement of Illorsuit—came into view.

# Kalaallit Nunaat Rockwell Kent-ip kingornagut: Illorsuit 1931-mi 2016-milu

*Piniartoq utoqqalisoq isaaffissuakkut sapangarasaartukkut appakaannermini Naalagarlu tikillugu oqamivoq: 'Naalagaq ataasiinnarmik apeqqutissaqarpunga: Sooq Kalaallit Nunaat ippernaqartippiuk?' Naalakkallu piniartoq utoqqaq qungujulluni saappaa, 'Kalaallit Nunaata nuna iluartoq unammillissanngimmagu.'*

**Illorsuit 1931 aamma 2016**

Illorsuit tungaanut timmisartornerup persuarsiornartup siullermeerlunga tikinninni Kalaallit Nunaata timmisartuutileqatigiiffiata qulimiguulianik Bell 212-mik timmisartunullu tallimanut pissutissaqartilluarpaanga ulluni pingasuni ingerlaarninni Straton-imiik Air National Guard-ip timmisartuanut New York-imiik Kangerlussuup Ilulissat Qaarsut Uummannap apuuffissamalu Illorsuit tikinnissaanut. Nukappiaqatut ilillunga mamakujuttuerniarfimmut qilanaangaarlunga appakaattuusivunga qulimiguulimmut ilaallunga akunnerup affaani sivisussusilimmik timmisartorninni igalaalimmi issiarusulersillunga. Qangattarpugut, Uummannallu qeqertaa kaajallattuuseriarlugu avannamut aallaannaqaagut Uummannami kangerluit

*Its seagirt isolation along with the grandeur of its stark, snow-covered tableland and higher peaks, the dark cliff barrier that forms its western shore—there is a glamour of imponderable mystery about the island, which dignifies it even at the gateway of a region of stupendous grandeur. Its cliffs, proclaiming inaccessibility, preclude the thought of human settlements.*[1]

As we came around the headlands into Illorsuit on the northeast end of the island, the sweeping beach of Illorsuit came into view. Colorful homes of blue, red, yellow, and green bordered the beach and fjord. The settlement, population 70, occupies a sloping, narrow spot of land between the Illorsuit Sound and the rugged black mountain walls that surround the community. My colleague called the settlement claustrophobic, as homes nestle into the crescent homeland of Illorsuit. With the surrounding wilderness, the only way out is by boat or snowmobile via the sound, or by helicopter on Mondays or Wednesdays, weather permitting.

We had landed some 300 miles north of the Arctic Circle on the barren headlands that lack any structures. The helipad was a flat dirt surface surrounded by orange barrels. Roughly 100 yards to the east of the helipad was the cemetery, and the town dump was 400 yards to the south. Town managers Jarte Korneliussen and Kasper Korneliussen met our helicopter, with Jarte leading us to our sleeping quarters for the next 10 days. My colleagues Axel Jeremiassen, Susan Vanek, Jette Rygaard, and I took up residence in the community center's exercise room and kitchen. Conveniently, the building also housed showers and the settlement's only running water. This first trip to Illorsuit was followed by two other extended visits that lasted on and off from June 2016 until March 2017—with our last trip ending during a snowstorm that left us trudging through deep drifts to get to the helicopter.

A very different experience was had by Rockwell Kent in July 1931 as he entered Illorsuit, which at the time was a remote settlement of fewer than 150 hunters and their families. Motoring in a small wooden vessel, Kent described, "nine interminable hours of our trip [to Illorsuit] from Uummannaq"[2] into the Illorsuit Sound, and past the steep, inaccessible cliffs of eastern Ubekendt Island. These barren, treeless tablelands crowned by thick, rolling green grass seldom if ever had visitors. Kent described the locals' "cry going up" as female residents in brightly colored native costumes and men in seal skin pants and anorak tops emerged from the numerous little mounds of earth that dotted the settlement to greet the newcomers. Several warehouse buildings and the manager's grand home were the only wooden houses in Illorsuit.[3]

Most Greenlandic homes north of Nuuk and Sisimiut at the time were one-room sod/turf and stone structures with glass windows and wooden walls on the inside. Pavia Nielsen[4] recalled living in such a house with his parents, siblings, grandparents, and an aunt and uncle. In contrast, today's modern Greenlandic family is more mobile, leaving the extended family less prevalent. Grandparents, for instance, are satisfied to live in their own homes, and many children leave home to further their education in Nuuk or Denmark and do not return. Understandably, career opportunities for these young Greenlanders are more plentiful in Europe or even in Sisimiut or Nuuk. This is one reason that the population of Greenland is slowly, but noticeably, decreasing.

**The Photographs**

I spent more than three months in Illorsuit photographing and presenting photography workshops to children in the settlement with my colleagues. The workshops culminated in a community exhibition of exceptional student photographs hung along with Kent's photographs that he had made in Illorsuit. A lively, full house of young and old attended this event, which took place in the schoolhouse poetically built above what remained of the foundation of Kent's original wooden 1931 house in which he lived with Salamina and her daughter. (A few years later, Kent built a small sod/turf house next to the wood house, where he wrote *Salamina*). A fine buffet feast of muskox, reindeer, fish, Greenlandic salads, and cakes was provided by the people of the settlement. Conversations about Kent's photographs and curiosity about people in the photos resulted in delightful interactions with those in attendance.

ikertunersaat qulaallugu. Iluliarsuit qaalloreqisut assigiinngeqisunillu ilusilersugaasut isikkiviliullugit, sukkut tamaana tungujuaartunik qalipaasersukkat imarlu qaalluarpaluttoq qulaallugu ingerlaarpugut. Kangitsinni Kalaallit Nunaata nunavittaata qaqqarsui aputitarasaartut takussaapput sermersuarmik tunuliaqutsigaasut, taakkulu kujaleraat Qeqertarsuaq, tassaniittorlu nunaqarfik Illorsuit—saqqummeratannguarpoq.

*Immamit avatangerneqarsimallutik immikkoortitaasutut ajasoornartumillu qularutissaanatik, aputitarasaartut nuna nerriuviusatut manissortalik inngigissunillu qaqqarasaartut, kitaatungimikkut assiaqutaasatut innaaqqinnersallit—uuttorneqarsinnaanngiusartumik nalorninartulittullusooq qeqertaq tamaani isaaffinngortitaavoq. Innaaqqinnertaasa tikikkuminaannera nittarpaat, inunnit najugaqarfigineqarsinnaanera takorlooruminaassitillugu.*[1]

Illorsuit qulangerlugit avannaata kangiagut ingerlaarnitsinni sissaa sioraattaajusoq takuneqarsinnaalerpoq. Illui qalipaatigissut, aappaaluttut, kajortut qorsuillu sissap qulaatunginnguaniipput. Nunaqarfik 70-it missinginik innuttaqartoq, sivinngarissumi annertungisassaanngittumi nunaminertami Illorsuit kangerluanni qaqqarsuarni qernertuinnaanerusuni ungaluneqarsimavoq. Angalaqatima nunaqarfik mattusimasutut taavaa, illunngui amilimaartutut ataatsimoorlutik inissisimanerat pillugu. Inuilaami pinngortitami avatangerneqarsimangamik, taamaallaat umiatsiamik qamuteralannilluunniit kangerluttaatigut tikinneqarsinnaapput, imaluunniit qulimiguulimmik ataasinngornikkut pingasunngornikkulluunniit, sila apeqqutaatillugu.

Qaasuittup killeqarfianit miilinik 300-nik avannarpasinnerusumi naggorluttumi amingakuluttumillu inuiaqatigiittut aaqqissuussaanilimmi inissisimapput. Miffigisarput sioraannaavoq nappartanik manngertornisunik ungalulersornneqarsimalluni. Miiterit 100 missaanik ungasissusilimmi iliveqarfik kangiatungaaniilluni inissisimavoq, eqqaavissuaqarfillu kujalliutsiarluni 400 miiterinik. Nunaqarfiup atorfilittai Jarte Korneliussen Kasper Korneliussen-ilu tikilluaqqusisuupput, Jartellu ulluni qulini tulliuttuni najugaqarfissatsinnut ingerlaqatigivaatigut. Suleqatikka Axel Jeremiassen, Susan Vanek kiisalu Jette Rygaard uangalu taakkuulluta nunaqarfimmi innuttaasut illuannut sungiusartarfiutigisumut igaffilimmullu inissippugut. Qujanaqisumillu najugaqarfipput innuttaasunut uffartarfiullunilu errorsisarfiuvoq, nunaqarfimmi qipiinnarissamik imeqartutuaalluni. Illorsuarnut angalanerput siulleq taanna piffissamik sivittuinertalinnik allanik marlunnik ilaqarpoq juunimi 2016-mi martsimilu 2017-mi—angalaneq kingulleq perluttalimmik ersinngimmik maangarnaqisumillu mittarfiliarnermi puigunaattorsiffiuvoq.

Misigisarlu allarluinnaaqqoorpoq Rockwell Kent-ip 1931-mi Illorsuarnut apuunnermini misigisimasaa, taamanimi asimioqarfiuvoq avingarusimalluni 150-ngajannik piniartuinnangajannik najorneqartoq. Pujortuleeqqamik qisuusumik angalanermini Kent eqqaamasalersaarluni allappoq, “angalaneq naassaanngiusartoq Uummannamiik akunnernik qulaaluanik sivisussuseqarpoq (Illorsuarnut),”[2] Illorsuit kangerluannut, persuanakululluni innaaqqinnerillu akulluqquallugit asimioqarfiup qeqertaanukarnermini. Nuna tamanna naggorluttoq, orpeqarani manissuajaartaqartoqarlunilu qorsorasaartarpoq tikinneqartarsimassaguni. Kent-ip inuttai eqqartoramigit arnartai qalipaatigissaartunik assukkersorlutik ‘qiammiartorlutik qummukartarput’-mik allaaserai angutitaallu puisit amiinik qarlillutik annorasserlutik naggorinnersamininnguisigut toorninnguatut tikilluaqqusiartortarnerarlugit. Quersuaq pisiniarfillu niuertorutsillu illui kisimik qisunnik sanaajupput sinnerilu issunik ujaqqanillu qarmagaallutik.[3]

Nuummiik Sisimiuniillu avannarpasinnerusumi taamani illut amerlanersaat issunik ujaqqanillu qarmangaasut ataasiinnarmik initaqartarput, igalaaminernik igalaaqarlutik iigaallu iluisigut qisuttalersugaallutik. Pavia Nielsen-ip[4] immiunneqarluni eqqaamasalersaarnermini eqqaavaa naliginnaasumik taamani angajoqqaami qatanngutimilu saniasigut, aanani ilaannilu attaat akkaallu illoqatigineqartarsimasut. Ullumikkulli Kalaallit Nunaanni killoq ilaqutariit nuttarnerupput, angajoqqaat qimanneqartarneri naliginnaaneruvoq. Aataakkut aanaakkut, assersuutigissagaanni, namminerisaminnik illoqarumanerusarput; qitornallu amerlasuut angerlarsimaffik perorfigisartik qimattarpaat Nuummut Danmarkimulluunniit ilinniariarlutik uteqqissanatik. Paasinartumillumi inuusuttunut suliffeqarnermikkut qaffakkiartorumasunut Europa, Sisimiut Nuulluunniit orniginarnerutinneqartarput. Tamatumunngalu pissutaaqataavoq Kalaallit Nunaata arriittumik, qularnaattumilli, inukilliartorusaarnera.

The initial inspiration for this project came from the collection of hand-colored lantern slides in the Rockwell Kent Collection at the Plattsburgh State Art Museum of the State University of New York at Plattsburgh. A lantern slide is a black and white transparency on glass that often was hand colored. The lantern refers to the projection system for the positive/transparency slide. In the 19th- and 20th centuries, these slides had great impact on audiences who attended travel and educational lectures, especially in visual disciplines. The lantern slides, like precious jewels, were stored in small wooden boxes made with separate slots to hold each slide. They became obsolete with the introduction of Kodachrome slide film in 1935.

Kent made many photographs of his travels and adventures as snapshots, recordings, references for artworks, or remembrances of his adventures and friends. Although his photographs served little as art for Kent, the photographs from his Greenlandic subjects were used to make the lantern slides. These colorful adventure capsules were Kent's visual entertainment for his publicity tours of the United States. A skilled orator, he spoke glowingly of the virtues of Greenland to sold-out audiences; since Kent first screened his lantern slides during his ambitious 47-city 1933 lecture tour, few have seen them. Along with his woodcuts, paintings, and etchings, they are essential documents of Kent's time in western and northern Greenland, providing a rare, hand-colored photographic record of areas that were minimally photographed by others. In addition, 90 percent of these lantern slides were never published.

Photographing Illorsuit and its residents in 2016 proved to be a slower process than I had anticipated, one that began with gaining the locals' trust, written permissions, and willingness to be part of the documentary project. Invitations to birthdays and special event *kaffemiks* became a great opportunity to meet people, eat great food, and eventually achieve this trust. These kaffemiks were parties and social gatherings with many cakes, coffee, fish, and game served to guests. Ultimately, I would photograph almost everyone in the settlement.

**A Photographic Methodology**

Though Kent's lantern slides were made for presentation to the American public in a personalized, somewhat romantic adventure story, the content of his photographs shows an intimacy with the Inuit people. His two years of living with Inuit communities gave him a unique acceptance of, insight into, and understanding of their culture and livelihoods. His photographs capture an artistic view of space and the Inuit's place and scale reflected in the environment.

Intent is an integral aspect of documentary and ethnographic photography. Kent's photographs were created for several reasons: as a reference for future paintings; as a representation of Inuit life; as the descriptive media for his lecture tour; and even as a memory record of his experience in Greenland. His intentions were not scientific, although his pictures provide a wealth of factual information about Inuit culture in the early 1930s. Kent created an ethnographic document that was in fact used as more of an entertainment device than as an anthropological study.

By synthesizing photographic methods, I provided documentary evidence of contemporary Greenland's society, culture, and environment through a visual study and exploration of its primal and cultural landscapes as well as its indigenous people. This is the method from which Kent's paintings and prints evolved. Many of his drawings and prints list captions and the names of his subjects. The lantern slides that he made, principally in Illorsuit, provide a wealth of information, especially combined with his writing in the *Greenland Journal* and *Salamina* books.

Sarah Pink wrote: "Fundamental to understanding the significance of the visual in ethnographic work is a reflexive appreciation of how such elements combine to produce visual meaning and ethnographic knowledge."[5] My comparative study project strove to do just this, as I examined both the artist and the communities at the time of production and how these images are engaged with and understood today. The goal of my photographs was to develop a reflective ethnography to determine reality vs. representation. A meaningful collaboration between study

## Assilisat

Qaammatit pingasut sinnitsiarlugit Illorsuarni assiliinermik assinillu saqqummersitsinermut suliaqartitsisarluta nunaqqatigiinnut meeqqanullu suleqatikkalu atorpagut. Sammisaqartitsinitsinni naggataarutigalugu nunaqarfimmi saqqummersitsisarpugut meeqqat assilisaat piukkunnarnersiorlugit nivinngaatitertarlugit Kentip assilisarisimasaanut sanilliussassatut. Atuarfik taanna qanga Kent-ip, Saalamiinap paniatalu 1931-mi Illorsuarni ukiitilluni najugaqarfigisaata qisuusup toqqaviani napavoq, tassanilu inuusuttut utoqqaallu uummaarissumik saqqummersitsinermi peqataalluarput. (Ukiualuit qaanngiunnerini Kent issunik ujaqqanillu qarmakkamik illumi eqqaani nappakkamini atuakkiani *Salamina* allappaa). Suliatsinnik saqqummersitsinitsinni nunaqqatigiit umimmaap neqaanik, tuttuminernik, aalisakkanik, kalaallit salaatiliaannik kaaginillu sassaallerput. Peqataasut Kentip assilisarisimasai inuillu assiniittut ilisarisinnaasatik eqqaakulullutik oqaloqatigiissutigilluarpaat.

Assit tamakkua seqersittakkanit Rockwell Kentip katersugaataani University Art Museum of the State University of New York-imiittumeersuupput, illoqarfimmi Plattsburg-mi. Seqersitat suliaaqqaaramik tamakkerlutik qalipaateqanngillat kingornatigullu ilaat assaannarmik qalipaaserneqarsimallutik. Seqersitat taakkua suliaapput ajunngittussamik ammasumillu suleriaaseqarnissamut atatillugu suliniummut. 1900-kkut 2000-kkullu ingerlatilernerini seqersitat tamakkua isiginnaarianut tusarnaarianullu peqataasartunut mikinngittumik sunniuteqarsimapput ilinniutaallutik, pingaartumik takusassiaanermikkut. Assit seqersitat, erlinnartuutitut, karsinut qisuusunut iliorarneqartarput ittujaarneqarlutik ataasiakkaarlugit paarilluarumallugit. Taakkulu tassaalerput maanna atorunnaarsimalersunit Kodachromet fiilmiinut seqersittakkanut 1935-mi nassaarineqartunut suliaasimasutut taaneqarsinnaasut ilaat.

Kent-i angalanermini assinik ikinngittunik assiliisaartarsimavoq tamatumalu saniatigut immiussisarsimalluni qalipaasarsimallunilu ikinngutiminik angalaqatigisartakkaminillu puigunaattorsiffigisarsimasaminik.

Assiliisaartarsimanermini Kent eqqumiittuliortorpalaartumik suleriaaseqarsimavoq, assimmi Kalaallit Nunaanneersut seqersittagassatut suliarineqartarsimapput. Amerikami naalagaaffeqatigiinni assit tamakku qalipaasersukkat aliikkusersuutitut angalaaqattaarluni takusassiarisarsimavai. Inini ulikkaarluinnarsimasuni oqalullaqqissutsi atorluartarsimavaa; aammalu Kent-ip illoqarfissuit 44-rut tamakkerlugit tikinniarsarinerini aattaat 1933-mi assit seqersittakkani qalipaasersortalersimavai, maannalu ikittuinnaat takusimavaat. Qisuit qiperukkat, qalipakkat kigartukkallu saniasigut Kent-ip assilisarisimasai tamakkua uppernarsaataallutillu takussutissaapput Kalaallit Nunaanni inooqataasimaneranut, mikingaluarlutillu assaannarmik qalipaasersungaasimasut tamakku allanit assilineqarumasarsimapput. Ilaannarlugulu seqersitakkat 90%-ii siornatigut allanit takuneqarsimanngisaannarlutik.

2016-mi Illorsuarni innuttaasunik assiliinermi peqataaffigisartakkanni allanut naleqqillugu pilaarnarnerusimasutut misinnarpoq, nunaqqatigiit tatiginnilersinnissaat peqataatinniarnissaallu eqqarsaatigalugu aammalu akuersiteqqaartarnissaat suliniummut taamaattumut. Inuuissiortoornerit *kaffillernerillu* inunnik naapitsinissamik periarfissatsialaapput, nerilluarnissamut tatigineqalernissamullumi. *Kaffillernerit* tamakku tassaapput naapeqatigiinnerit nalliuttorsiornerit kaagillerfiulluartaqisut, kissartunik najoqqagassartallit, aalisakkanik, pisarisimasanillu allanik sutorluarfiusaqisut. Taamaattoqartillugulu qaasut sapinngisamik amerlanersaat assiliumasarpakka.

## Assiliisaarnermi periaaseq atorneqartoq

Naak Rockwell Kent-ip assilisarisarsimasai Amerikkami innuttaasunut alutornarsakkatut takusassiarineqarsimagaluartut, immaqalu nuannersunngorsakkatut oqaluttualiamut assersuunneqarsinnaallutik, assit tamakku imarisamikkut taamani inuit inooriaasiinik takussutissiisuupput. Ukiuni marlunni inooqataanermini kulturimik inooriaatsimillu asseqanngiusartumik takussutissiinissamik paasisaqarnissamillu periarfissaqarluarsimavoq. Assit taamanernitsat eqqumiittuliortorpalaartumik pissusiusimasunik takussutissiisuupput sumiiffinnik avatangiisinillu.

communities and photographer developed as I participated in daily life and built relationships that enabled a representative version of how reality is constructed. A collection of experiences that developed trust reduced the sense of the "other" and the "cultural distance" that Macintyre and MacKenzie's essay in *Anthropology and Photography* describes as problematic of ethnographic and photographic studies.[6]

Rockwell Kent's 1930s photographic documentation of the time he spent enchanted by Greenland is an invaluable ethnographic resource that provides a perfect backdrop for the present study. For this study, photography was both the visual recording medium and a visual research method. Photography is the paradigm of comparison. A comparative system of sequential, staging, representational, captioned, contextual information, and decontextualized photographs can be used as a comparative foundation.[7]

**Notes**

1 Rockwell Kent, *Salamina* (New York: Harcourt, Brace and Co., 1935), 6.

2 Kent, *Salamina*, 7.

3 Ibid., 6.

4 Pavia Nielsen (Uummannaq hunter and elder), interview with author, August 2016.

5 Sarah Pink, *Doing Visual Ethnography: Images, Media and Representation in Research* (London: Sage Publications, 2001), 29.

6 Martha Macintyre and Maureen MacKenzie, "Focal length as an analogue of cultural distance," in *Anthropology and Photography*, ed. E. Edwards (London: Yale University Press, 1992), 158.

7 Richard Whelan, *Double Take* (New York: Clarkson N. Potter, 1981), 15–16.

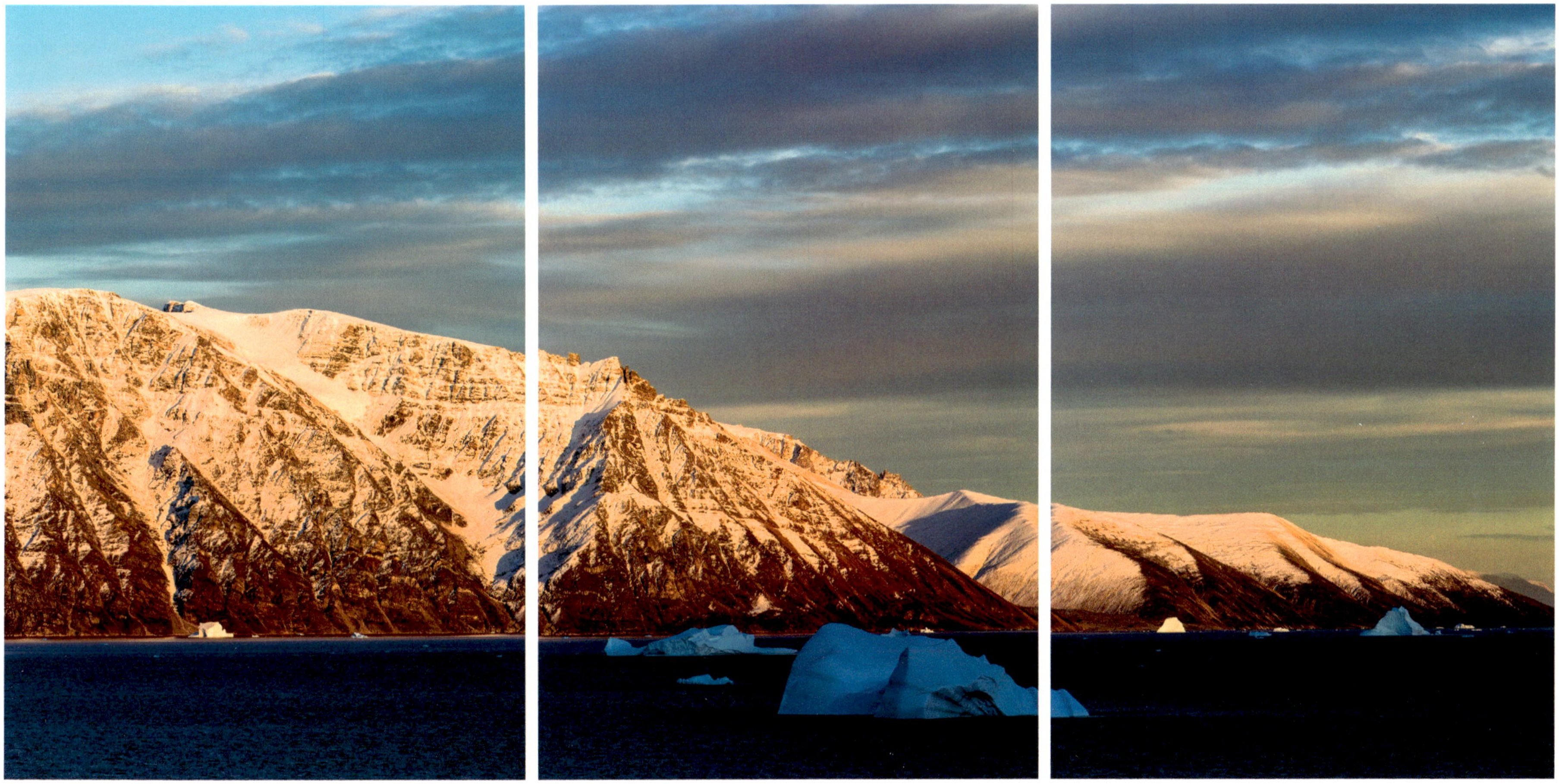

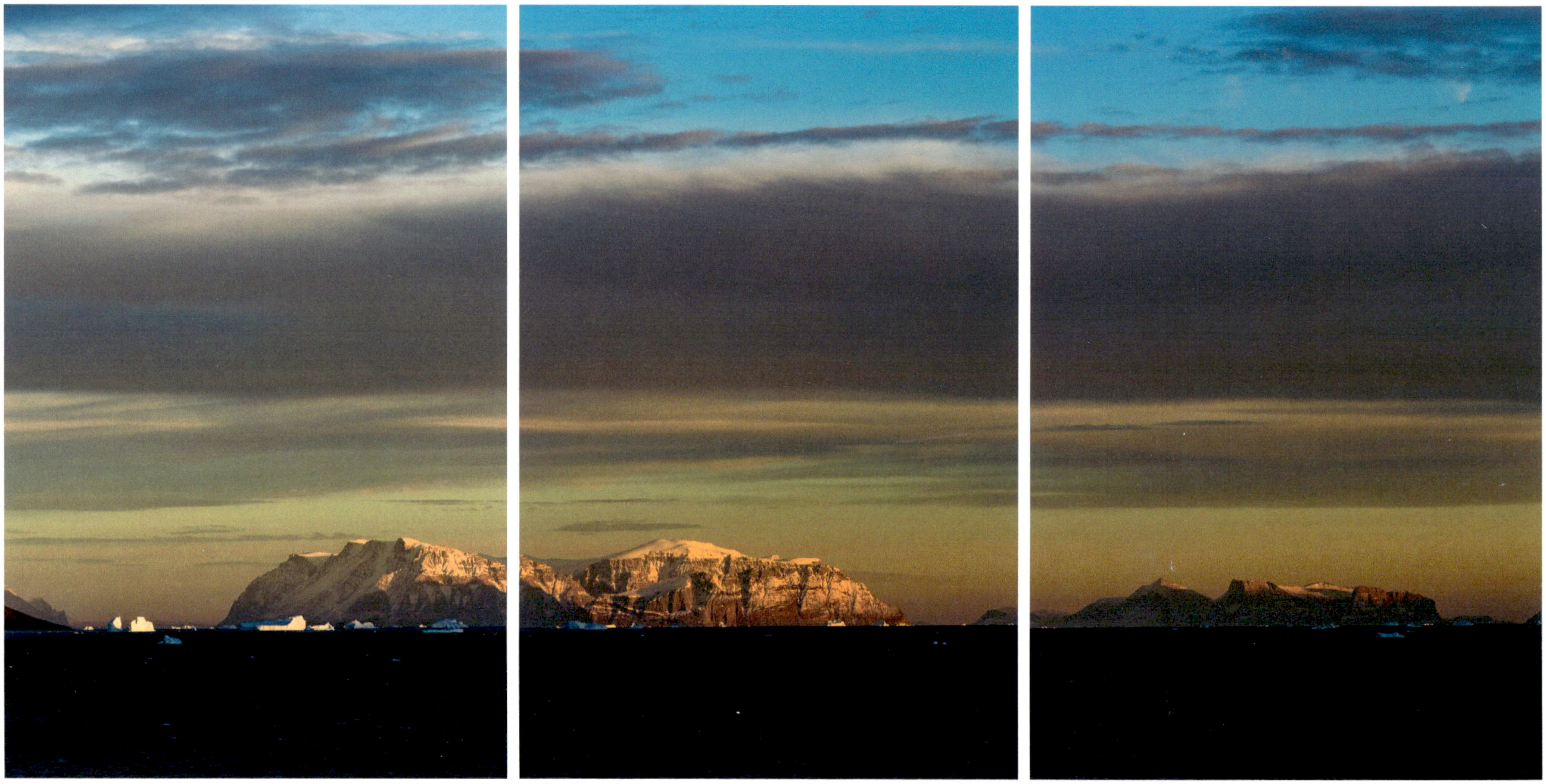

Assilisat uppernarsaataapput naggueqatigiit inooriaasiinik takussutissiisut immikkuullarissut. Kent-ip assilisarisimasai arlalinnik tunngaveqarluni assilisaasimapput; inuit inooriaasiannik takussutissatut, tusagassiorfitsigut saqqummersitsisarnerni takussutissiassatut, kiisalu inuunermini Kalaallit Nunaannissimaneranut eqqaassutissatut. Ilisimatusarnissamik siunertaqarsimanngilaq, naak taamani 1930-kkunni inuit kulturiannik takussutissartaqaraluartut. Kent-ip assilisarisimasai inooriaatsimut uppernarsaataapput aliikkutassiaanerusimasut inuiaat pissusaannut ilinniutissiatut atorneqarsimanerminngarnit.

Assileriiaatsinnut ullumikkut atorneqartartumut periaaserisannut tapertaliuttutut nalitsinni Kalaallit Nunaanni inuiaqatigiit, kulturiat avatangiisaallu atuartut peqatigalugit misissuisutut ilisimasassarsiortutullu takussutissiariniarsimavakka, inuiaat nunap inoqqaavisut naatsorsuutigalugit. Periaaseq tamanna Kent-ip qalipangarisarsimasaanik assiutaanillu nangitsineruvoq suliatut ingerlaqqittoq. Qalipagarpassuisa assiuterpassuisalu takussutissartaanik sammisarisimasaanillu nittarsaassisutut. Assit seqersittakkat suliarisimasai, amerlanersaat Illorsuarnit pisut, paasissutissaalluarput, pingaartumik atuakkiarisimasaanut marlunnut *Greenland Journal*-imut *Salamiinamullu* ilanngukkaanni.

Sarah Pink ima allappoq: "Tunngaviusumik naggueqatigiit inooriaasiannik takusat najoqqutaralugit suliarisartangaasut akuersaartumik oqaasertalissagaanni takusassiatut pilersiortukkat akuleriisillugit isumaat naggueqatigiissutsimik ilisimasaqalertoqarsinnaavoq."[5] Sanilliussillunga taamatut assiliisartutut suliaqarniarsimavunga, eqqumiittuliortutut nunaqqatigiit misissortarnerini taamani suliat inerneri imminnut atassuserneqarsinnaanersut aammalu nalitsinni qanoq paasineqarsinnaanersut. Assiliisartuuninni siunertaraara naggueqatigiissuseq ersersissallugu tassuunalu takussutissiaralugu piviusoq illuatungaanilu nammineq isigisaq. Siunertaqartumik nunaqqatigiinnik misissuineq assiliisarnerullu ineriartortinnissaa ulluinnarni inooqataanikkut piviusullu taamaattuussusaanik ilusilersueqataanikkut. Taamaalilluni misilittakkat katersornerisigut tatigeqatigiinnissamik ineriartortitsinikkut misigisartakkap "taakku" aammalu "kulturikkut nikingassutsit" Macintyrep Mackenziellu atuakkiarisaanni taaguutilimmi Anthropology and Photography ajornartorsiutitut taallugu allaaserisimmassuk naggueqatigiissutsimik assiliisarnermillu misissuinerminni.[6]

Rockwell Kent-ip 1930-kkunni assiliisartutut uppernarsaatiliaanni Kalaallit Nunaanni naggueqatigiissutsimik assigineqanngitsunik takussutissaqalersippaatigut nalitsinni ilinniagaqarniartuni periarfissatsialaasunik. Taamatut ilinniarniaraanni, tassa assiliisarnermik aammalu takusanik immiussatigut nutaarsiassalerinertigut isigisanillu misissuisarnerni periaaserineqartartutigut. Assiliisarneq pissusilersuutinik sanillersuussineruvoq. Sanillersuussineq aallarniutaasinnaavoq, takutitseqqaarnermi, nittarsaassinermi, nivinngagassnik allagarsiisarnerni, ataqatigiissaakkamik paasissutissiinermi oqaasertalersugaanngitsunillu assitalersuisarnerni sanillersuusseqqaarnermi tunngavissaasinnasutut atugassiatut.[7]

1 Rockwell Kent, *Salamina* (New York: Harcourt, Brace and co., 1935), 6.

2 Kent, *Salamina*, 7.

3 Ibid, 6.

4 Pavia Nielsen-ip immiunneqarnera, allaaserinnittup oqaloqateqarneratigut, aggustip 16-ani, 2016.

5 Sarah Pink, *Takusassiaq ataaseq etnografii atorlugu: Assit, mediallu atorlugit misissuinertigut nittarsaassat* (London: Sage Publications, 2001), 29.

6 Martha Macintyre and Maureen MacKenzie, 'Kiisalu taakkua saniasigut marluk assiliissutip isaata qeqqaniik analogitut kulturi assigiinngissusaat nittarniarlugu,' in E. Edwards, ed., *Pingasullu antropologi assiliinerlu tunngavigalugit* (London: Yale University Press, 1992), 158.

7 Richard Whelan, *Sisamallu assit qaleriittuusitillugit iluatsitaakkat* (New York: Clarkson N. Potter, 1981), 15–16.

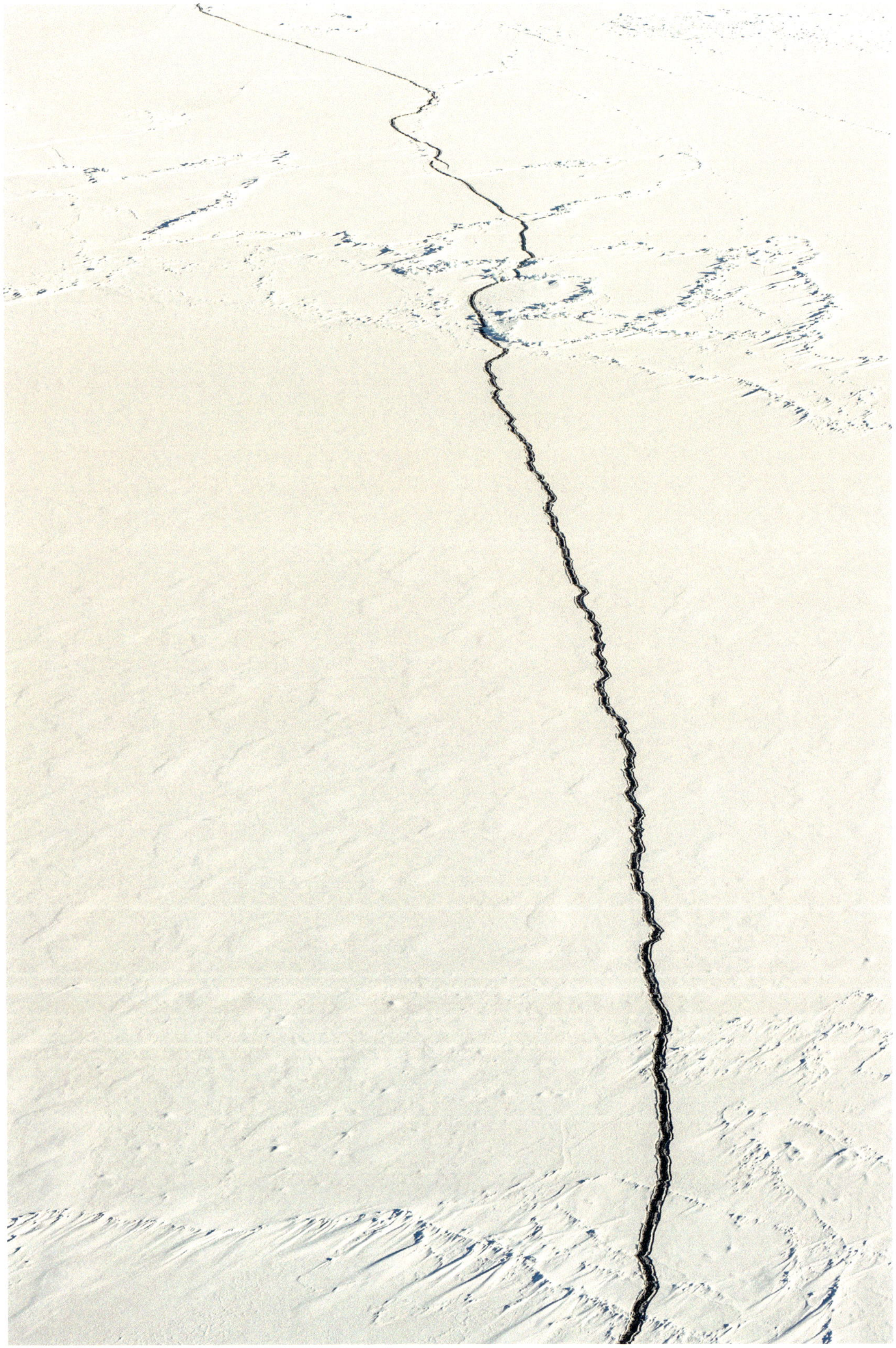

UUMMANNAQ DISTRICT 2017
Illorsuit Sound ice split makes for threatening snowmobiling and dog sledging. The dogs can sense these dangerous situations and thin ice.

# Uummannnaq District to Illorsuit

DENIS DEFIBAUGH

**The expansive, horseshoe-shaped, black sand beach on a pristine, clear fjord, bordered by majestic mountains and glaciers, invites a vision of an exotic tourist destination.**

But in the remote community of Illorsuit, Greenland, where there is no running water, "We are not that many here," resident Erik Qvist said in 2016. "We all know each other pretty well because there are so few of us."[1] Located on Ubekendt Island, Illorsuit is nearly 300 miles north of the Arctic Circle and roughly 60 miles from the main town of Uummannaq, in an extensive fjord area of west-central Greenland where glaciers line the jagged coast. It has, for many years, been a fishing community, with halibut its cash crop. Seal and whale are still hunted for food, for Illorsuit's people and their dogs. The small settlement's *Pilersuisoq*, or grocery store, at its center is the only place where residents can buy food, clothing, and other needed goods, albeit in limited quantities. Winter's arrival means that one will not see a ship with cargo and fresh produce from December until April, sometimes May. "We're used to living in such a place like this, so no big rants about it," Qvist affirmed. But Jarte Korneliussen, another resident, said in 2016, "When summer and the sun start to rise, Illorsuit livens up. We cook on the ice, play soccer and rounders, we do everything together—children, grown-ups, toddlers, and old people."[2] A year later, all this changed.

In mid-2017, a large rock slide into the Illorsuit Sound created a huge tidal wave that deposited sea water into more than half the town. Authorities evacuated the community and prohibited residents from returning due to the probability of more landslides.

Illorsuit's population numbered about 165 when Kent arrived; by 2016, it had shrunk by half. During Kent's time in Illorsuit, turf houses were small and families were large. Pavia Nielsen recalled living in a house made of sod and

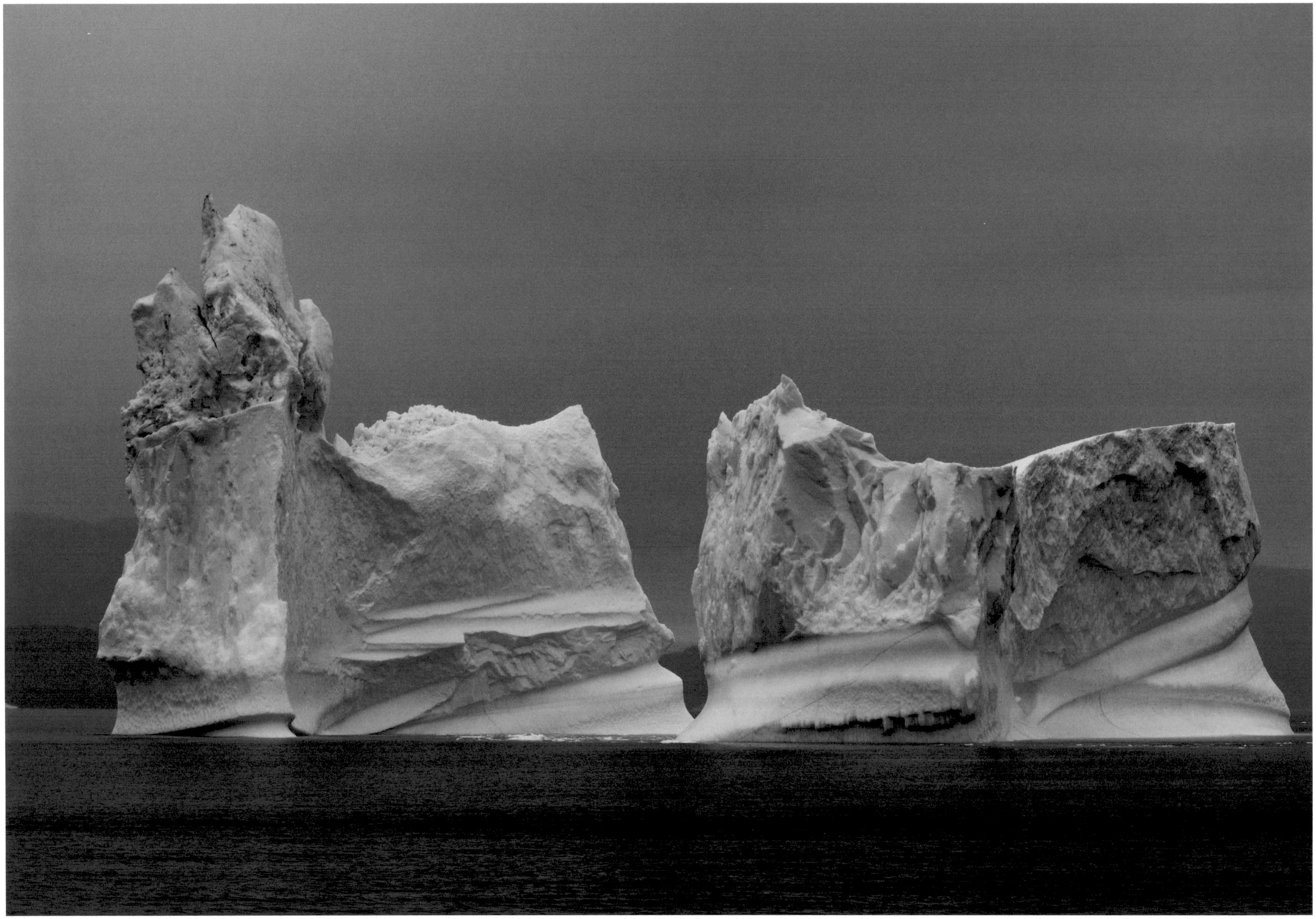

UUMMANNAQ FJORD 2016
Portrait of an iceberg.

stone with his siblings, parents, grandparents, and an aunt and uncle.[3] Today, such traditional turf houses are considered artifacts, having been replaced by wooden houses in the 1950s and 1960s. During the 1930s, however, extended nuclear families supported their children's and grandchildren's life experiences. Even while bringing up children, every skill supported the family's survival. Men fished and hunted, while women prepared seal hides and whale oil for income. For boys and girls, much of this traditional knowledge has been shared with subsequent generations.

While Illorsuit was Kent's ultimate destination in 1931, he arrived via Uummannaq, which was the district's shipping hub. Kent's supplies, including building materials, arrived shortly after he did; but much to his chagrin, noticeably missing were the precious and much-needed nails for the home he would eventually build.

Uummannaq was a scenic doorway into Illorsuit for Kent. Located in the middle of the sprawling Uummannaq Fjord, the town is dominated by the massive Mount Uummannaq. It may be one of the most beautiful locations in Greenland; the mountain rules the fjord and can be seen from Illorsuit. Homes line the barren rock and cliffs of the island; the harbor is the hub of the town's activities.

"Uummannaq is fantastic, with amazing views," notes resident Ann Andreasen. "Our iceberg moves every day and is a piece of abstract art. The Nuussuaq Peninsula—our Great Wall of China—keeps many people outside. But some people come through. And we are lucky to have those people visit us."[4]

**Notes**

1 Erik Qvist (Illorsuit resident), interview with author, September 2016.

2 Jarte Korneliussen (Illorsuit resident), interview with author, October 2016.

3 Pavia Nielsen (Uummannaq hunter and elder), interview with author, August 2016.

4 Ann Andreasen (director, Uummannaq Polar Institute), interview with author, September 2016.

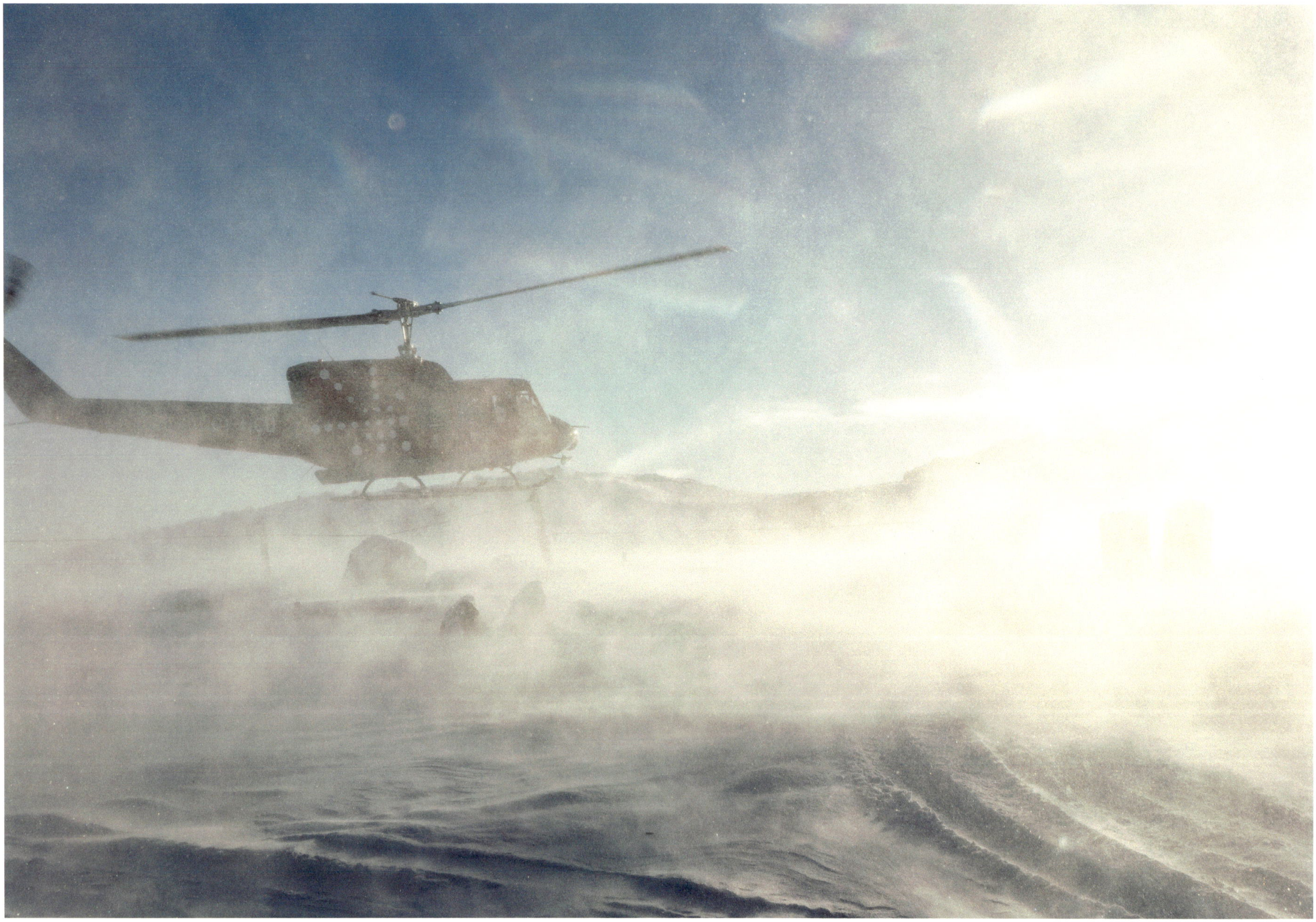

ILLORSUIT 2017
Air Greenland helicopter landing at the Illorsuit helipad. Flights hold nine passengers and are scheduled for Mondays and Wednesdays—weather permitting.

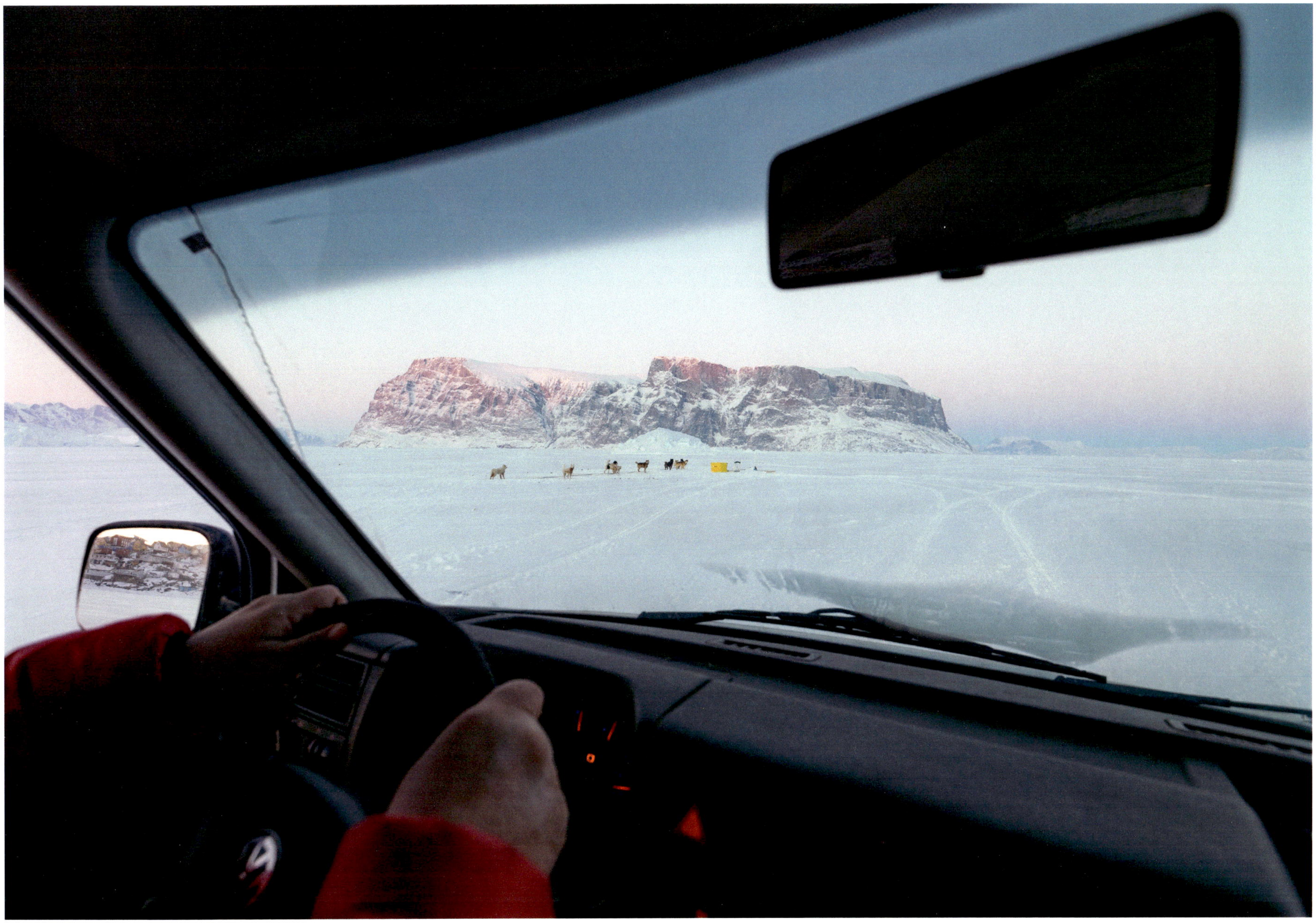

UUMMANNAQ 2017
Apartment manager Hans Mørch driving on the frozen fjord at sunset. Conditions were a little wet in places, so the drive was shortened.

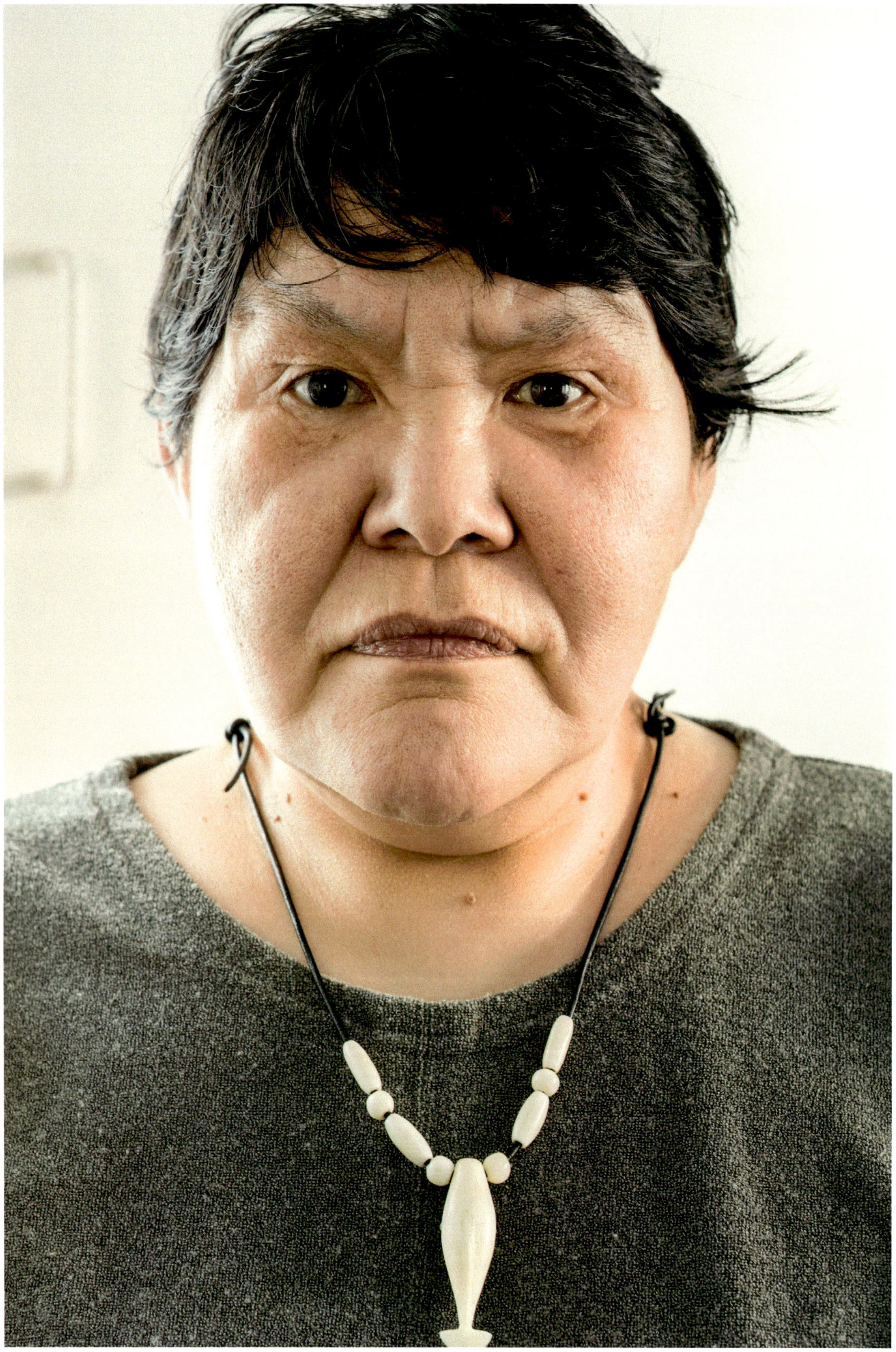

ILLORSUIT 2016
Ester Zeeb, midwife in Illorsuit. Zeeb is a well-known last name in Greenland; many of Kent's friends shared it.

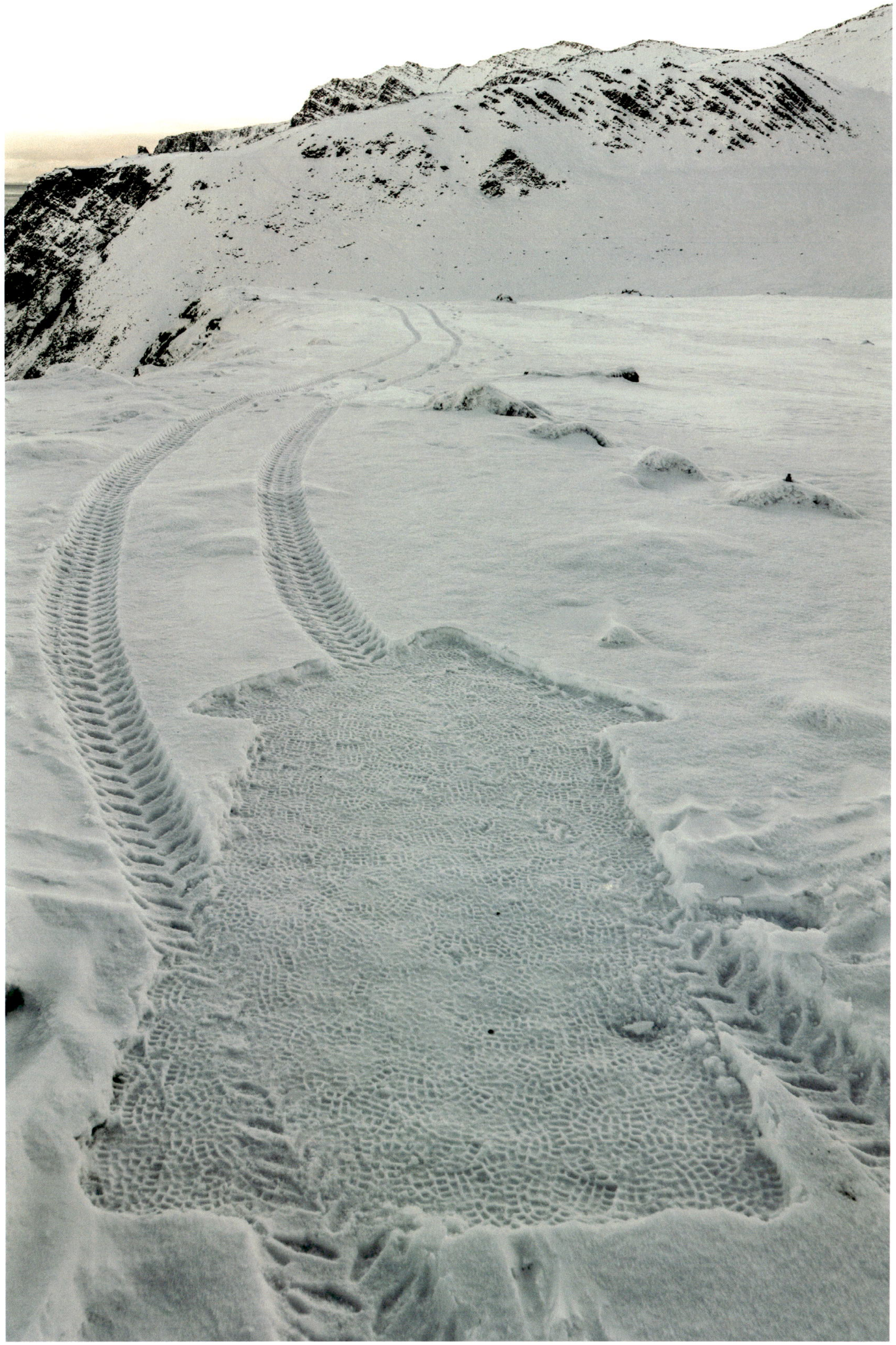

ILLORSUIT 2016
Mysterious arrow near the helipad.

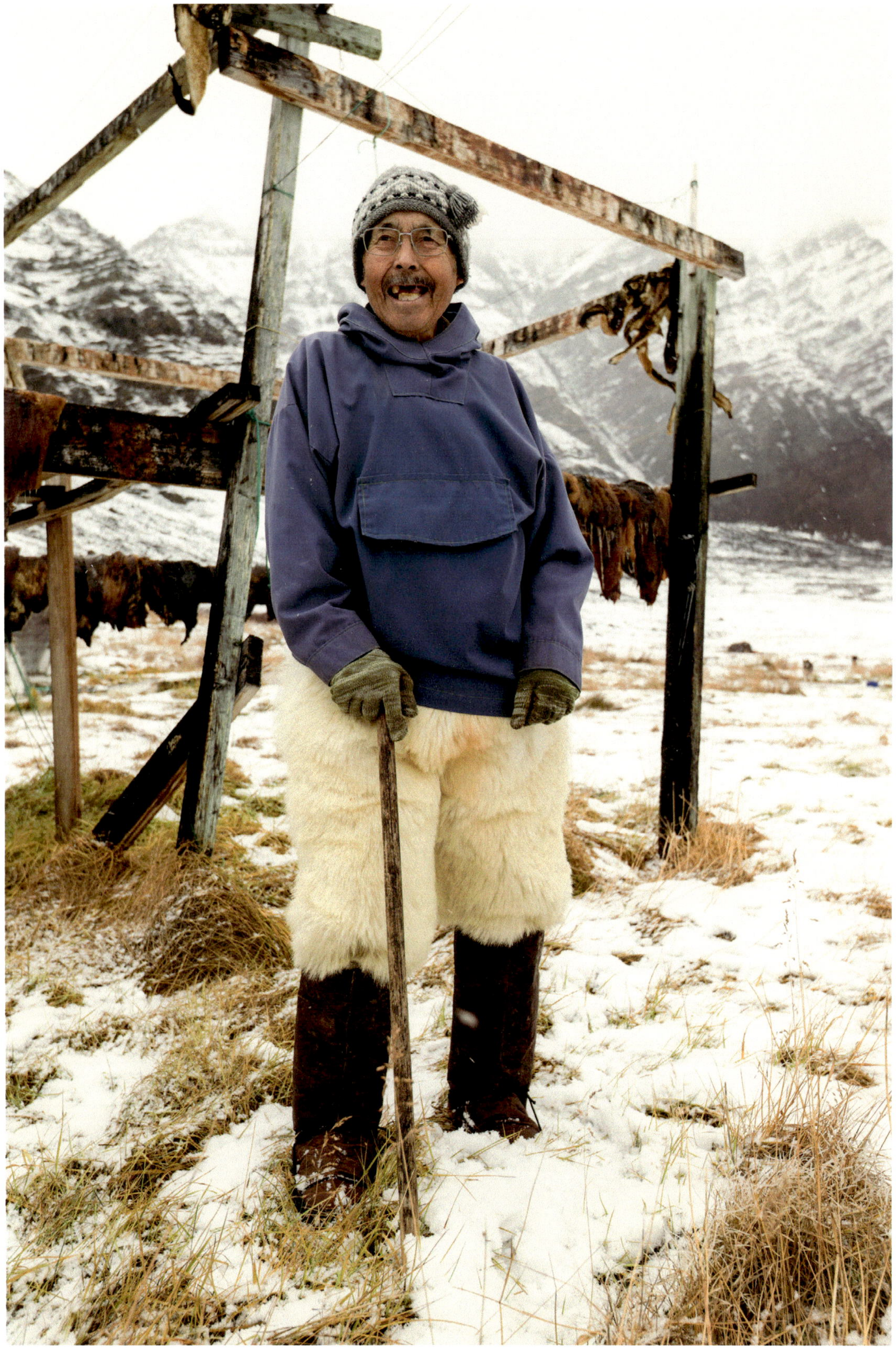

ILLORSUIT 2016
Aron Nielsen in traditional winter clothing: bear fur pants, anorak top, and seal skin boots.

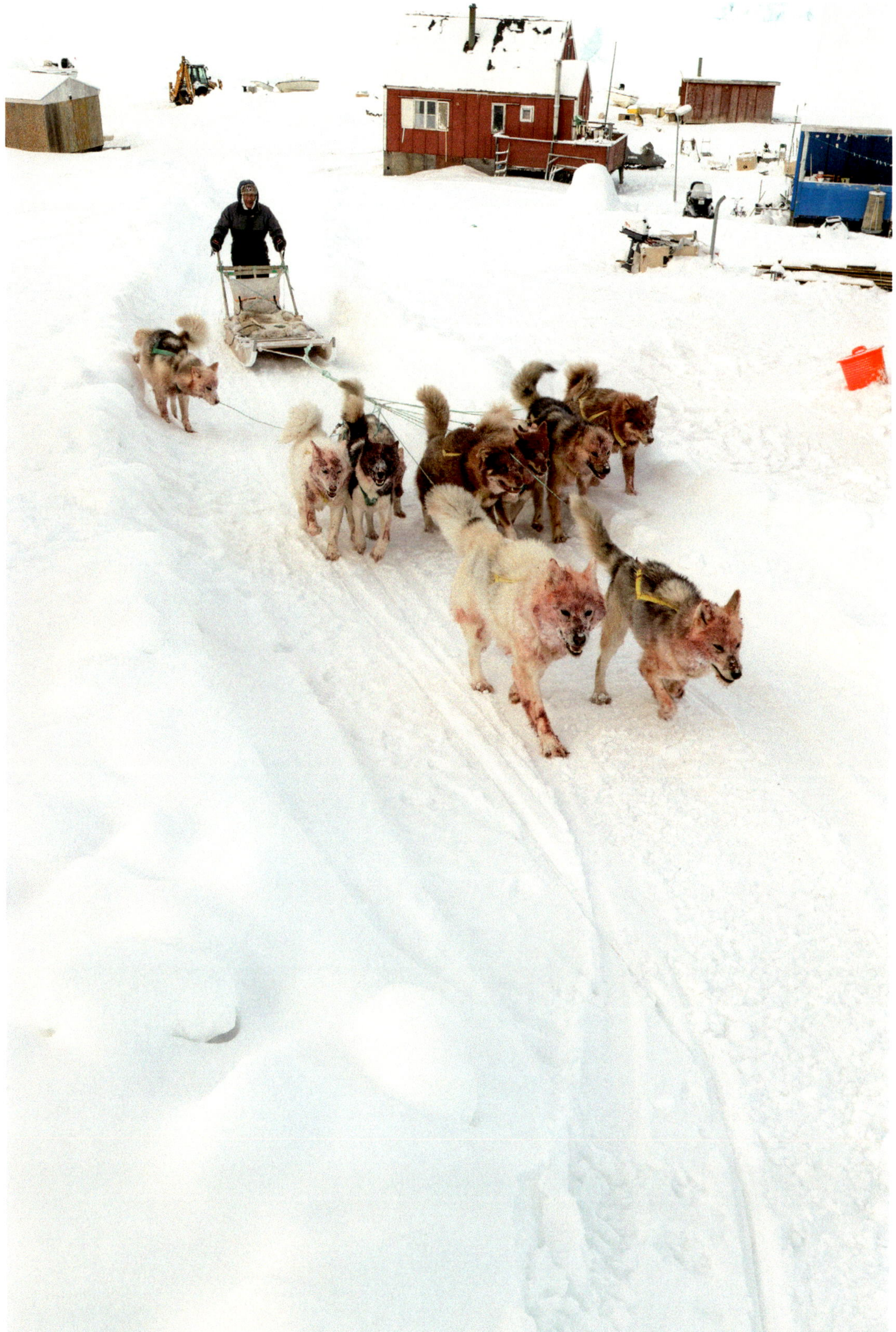

ILLORSUIT 2017

Hunter Erik Qvist returning from hunting on the frozen Illorsuit Sound. The dogs found some food during the hunt.

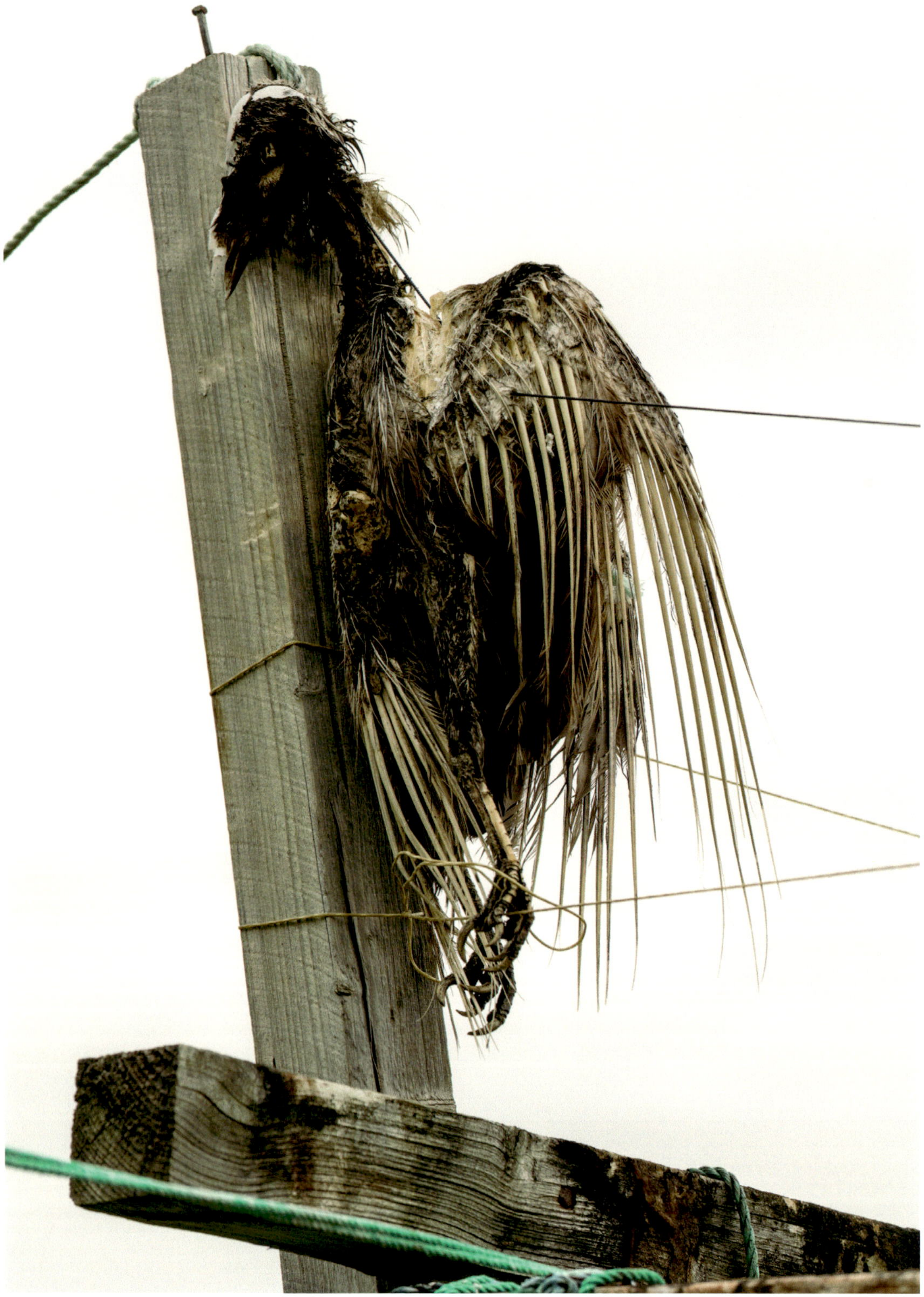

ILLORSUIT 2016
A raven carcass hanging on a drying rack. This is a warning for ravens and other birds to keep out.

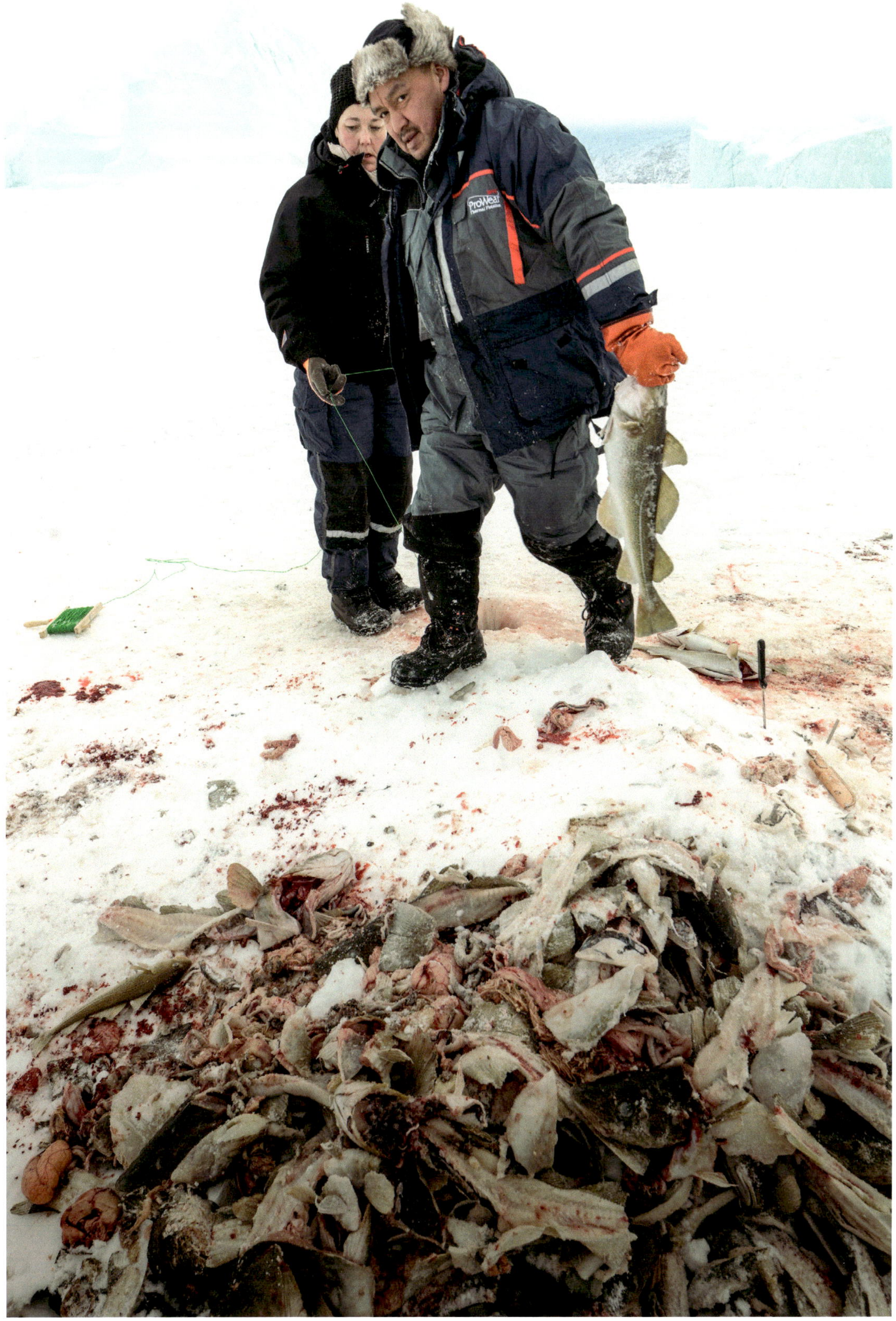

UUMMANNAQ 2017
Kristin and Jurgen Zeeb ice fishing for cod. Almost every day they are fishing and drying fish.

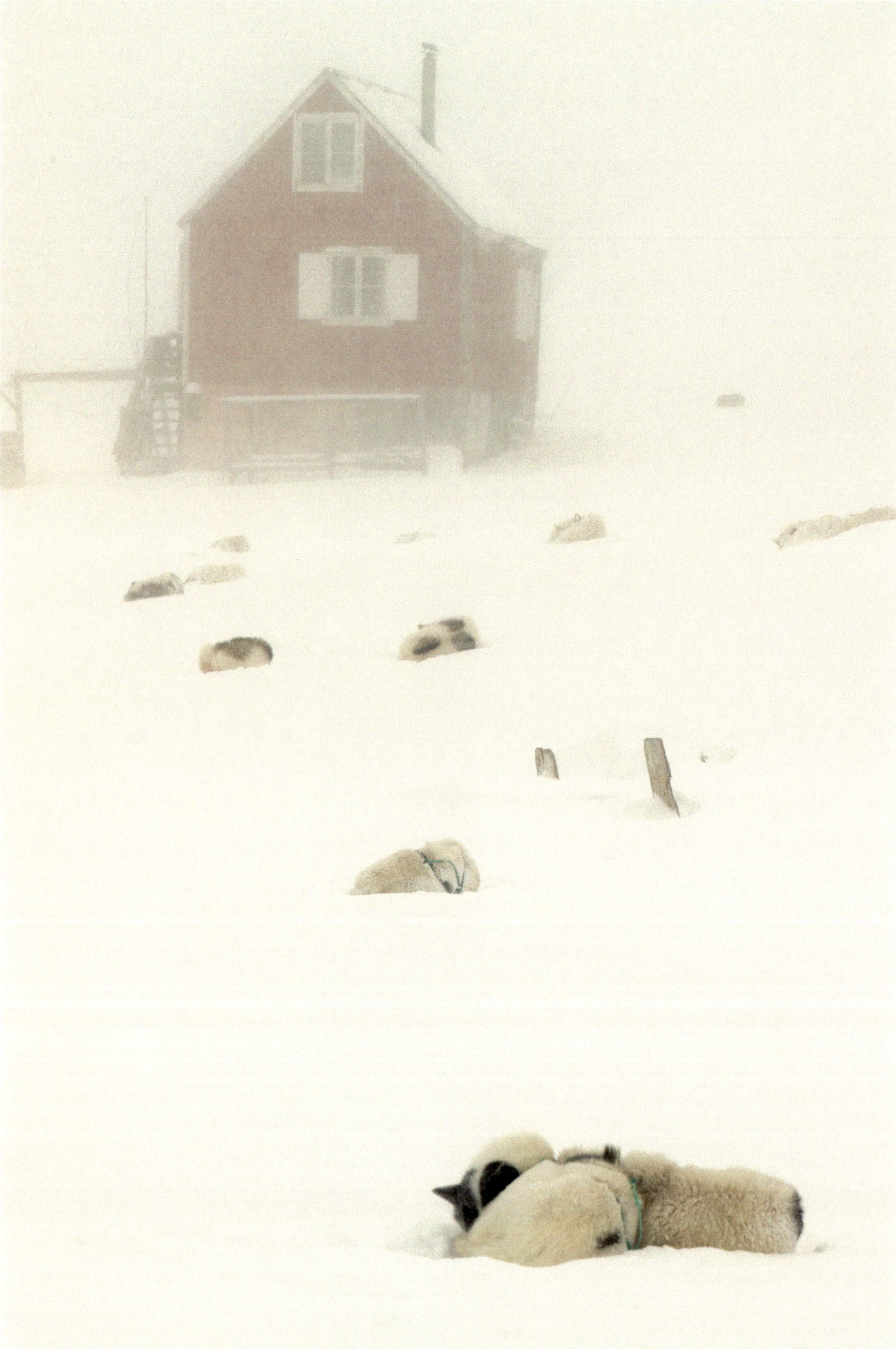

ILLORSUIT 2017

Dogs curled up for protection from a snowstorm. They may end up totally covered by snow; about two feet fell this day.

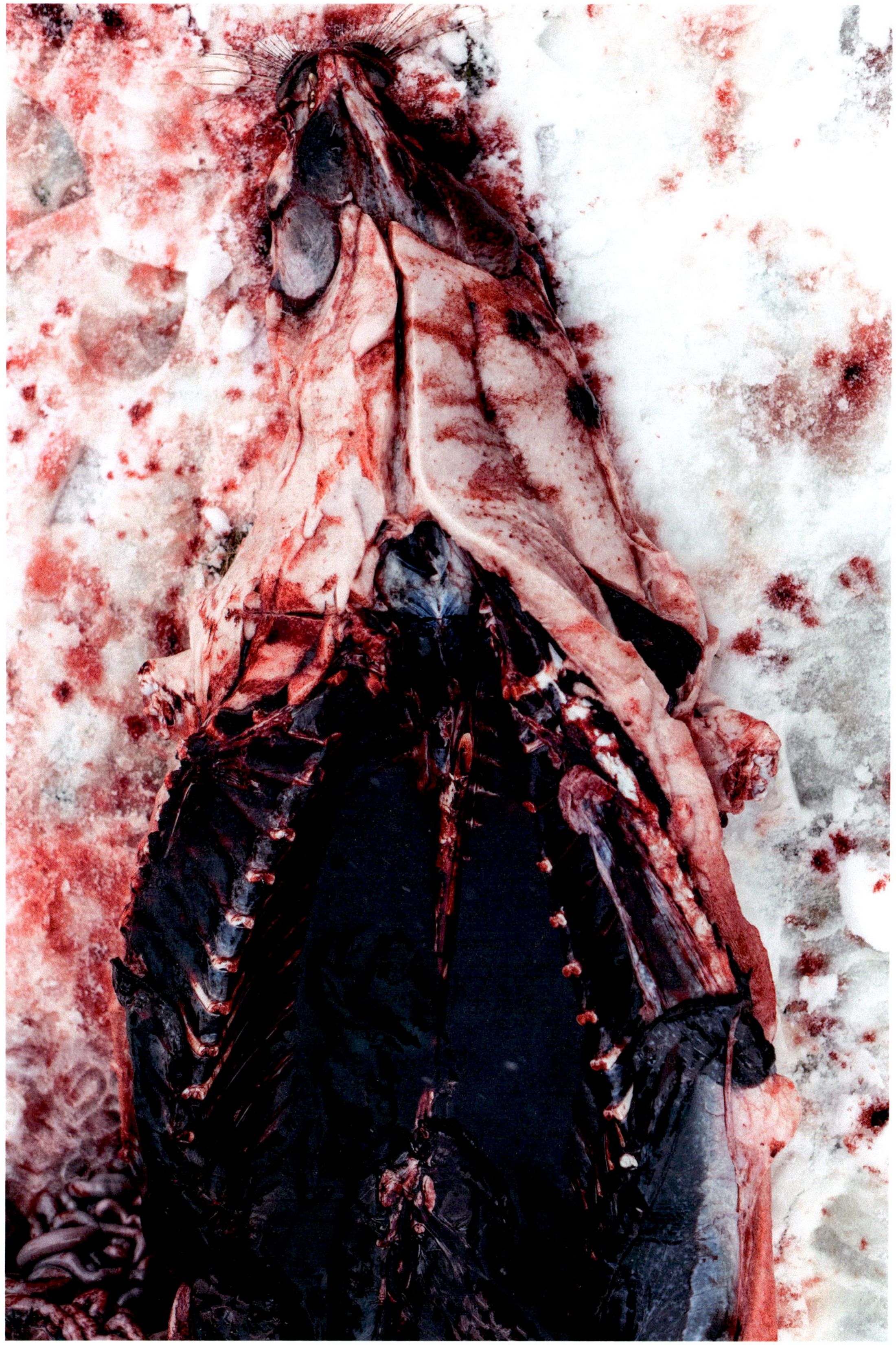

ILLORSUIT 2017
A skinned and gutted seal carcass. Blood is almost black. The meat is enjoyed by the people and is a treat for the dogs.

UUMMANNAQ 2017
The Jeremiassen family hike and play on the fjord during mid-winter.

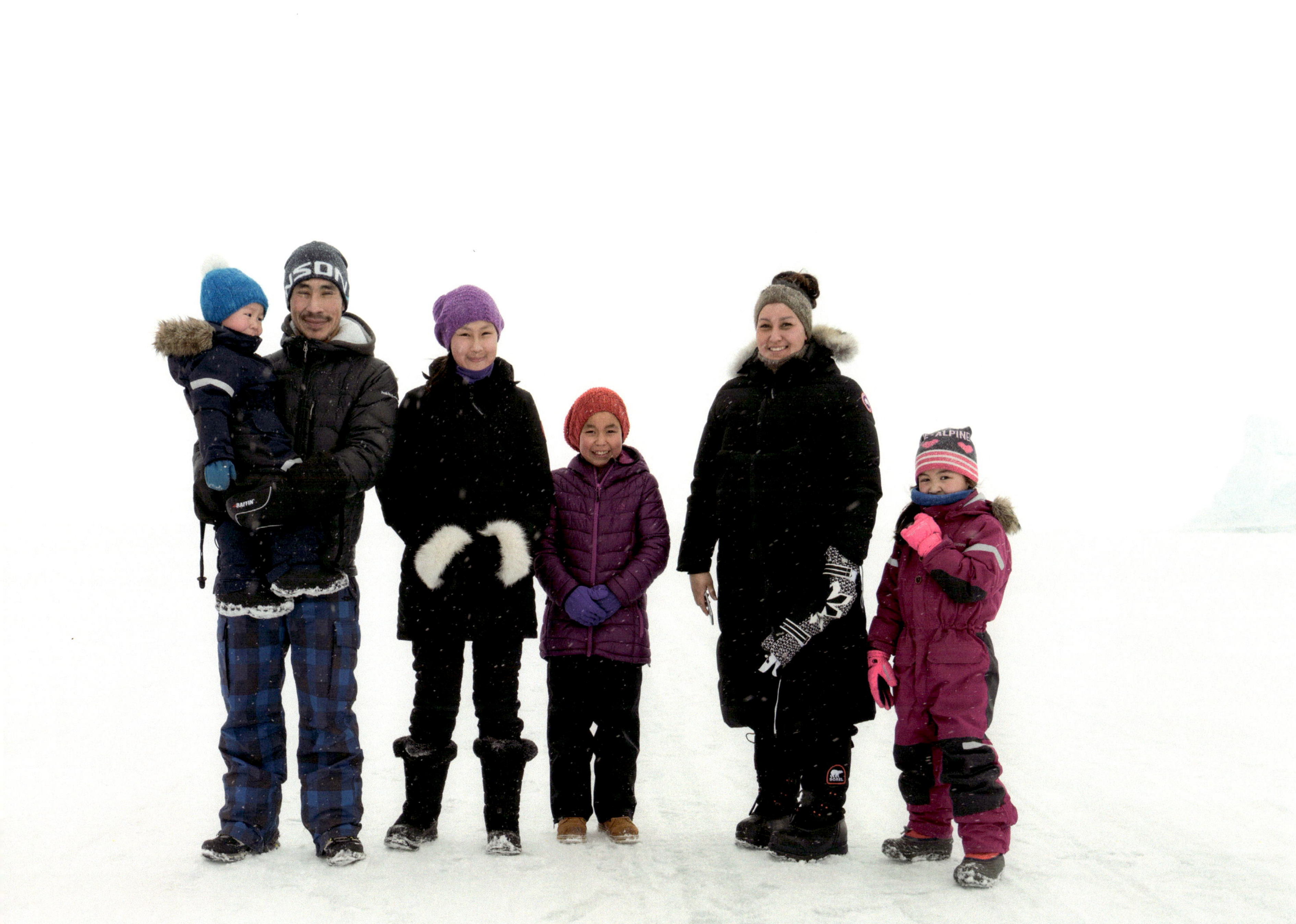
BAFFIN
ALPINE

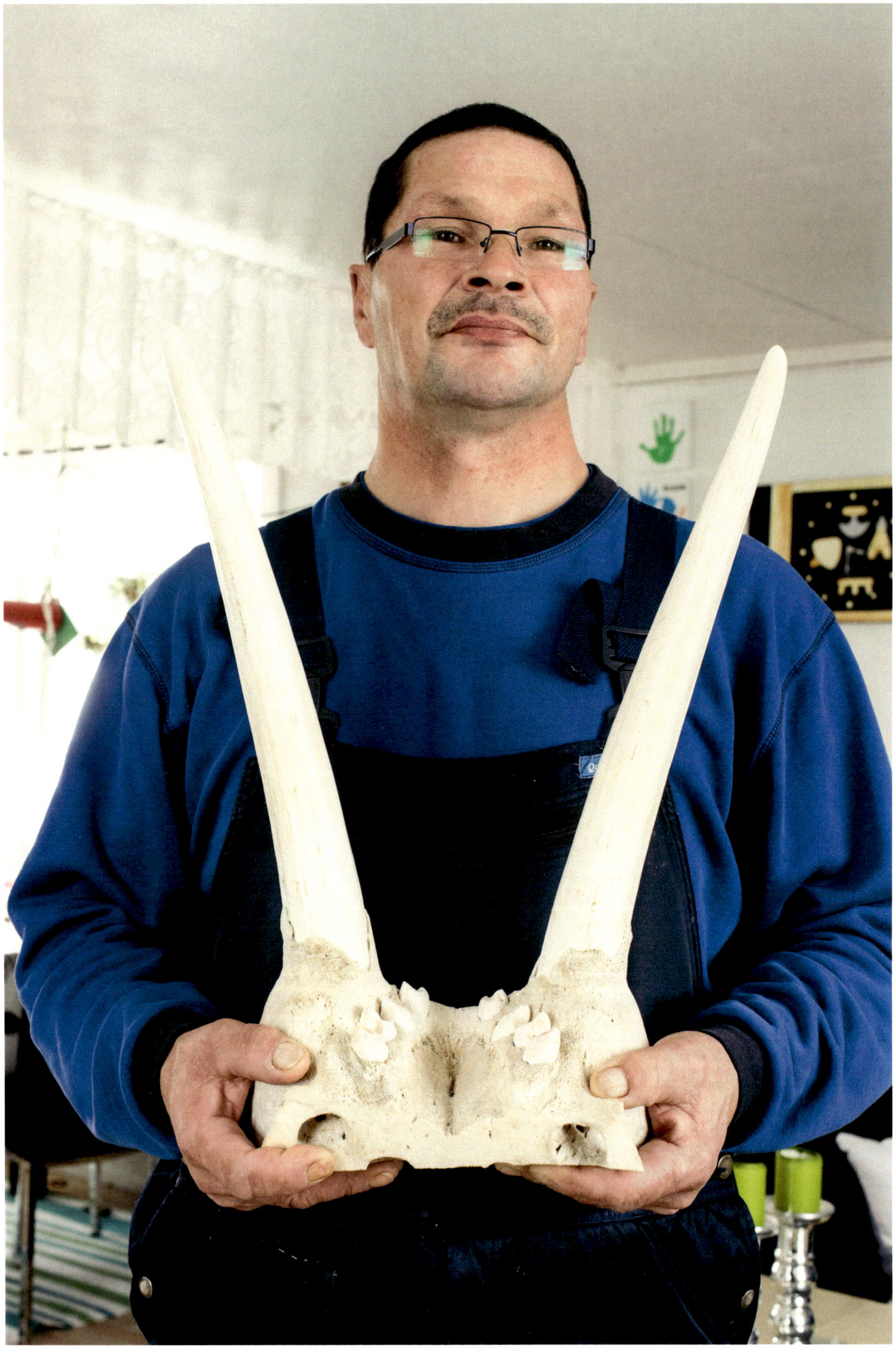

ILLORSUIT 2016
Aligot Ottosen holding one of several large walrus tusks from his trophy collection. He is known as a great hunter.

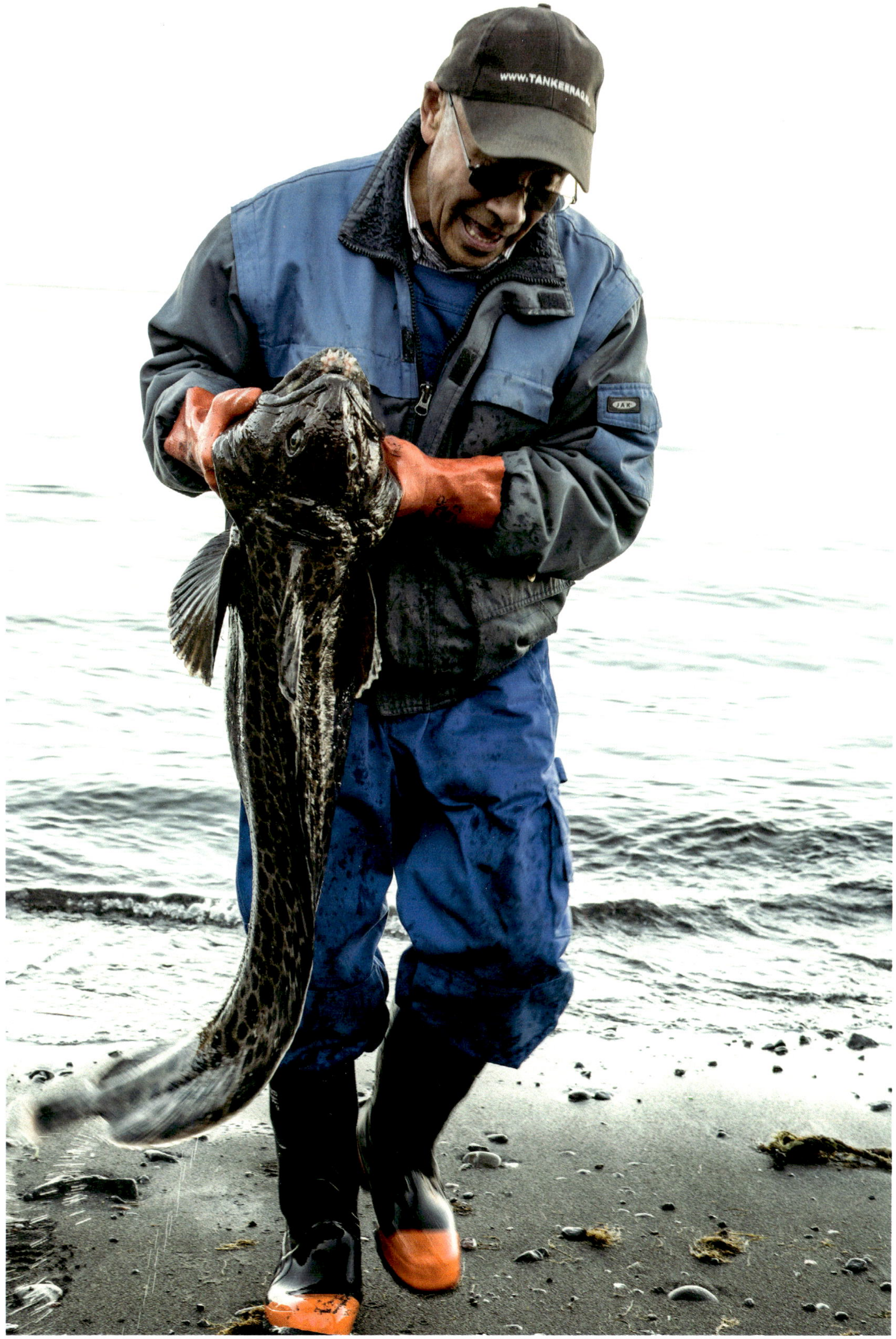

ILLORSUIT 2016

Benjamin Jonaghansen catching wolffish. After setting a 200-yard long-line with many baited hooks three hours earlier, he pulled in two wheelbarrows full of fish.

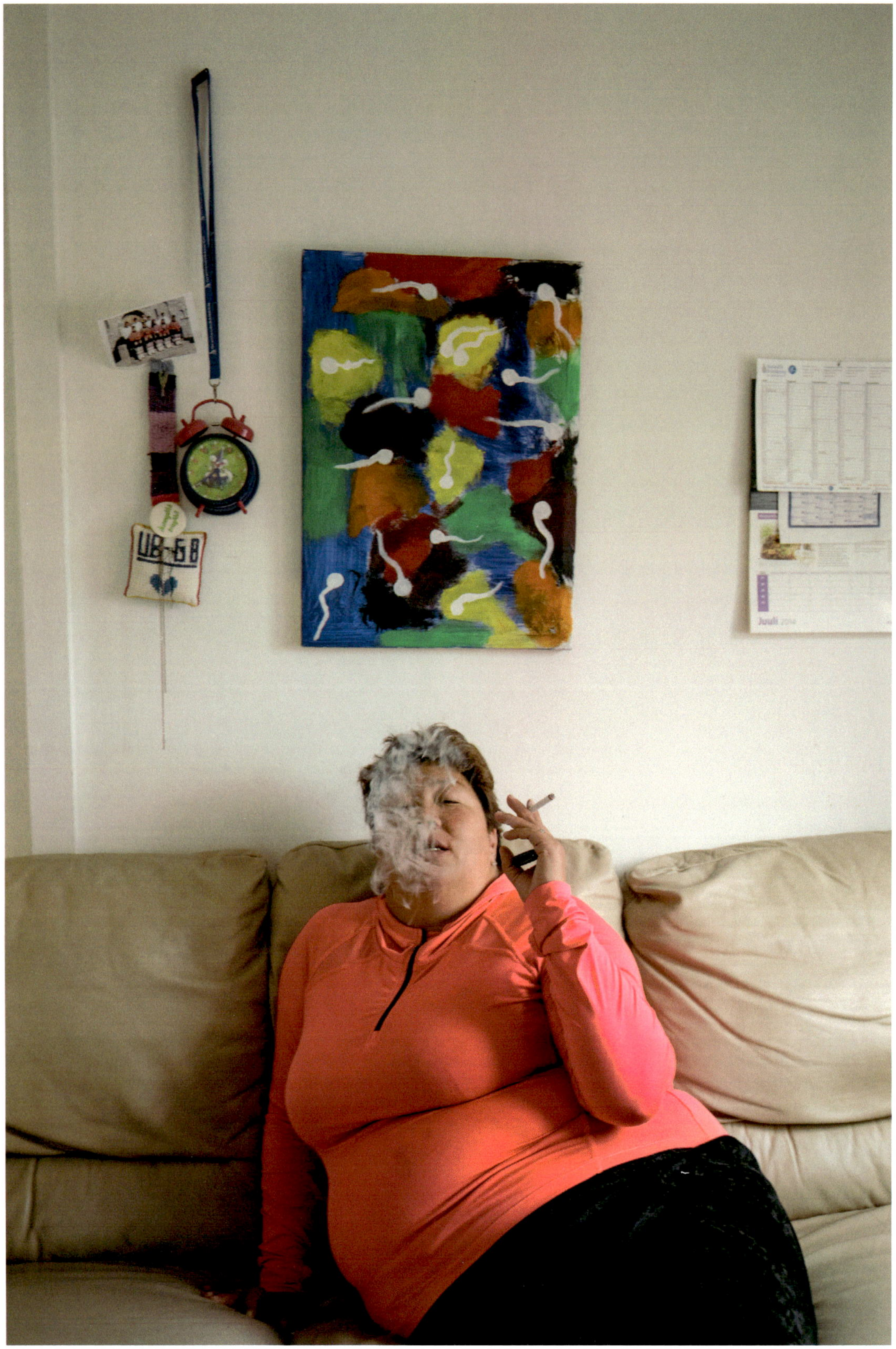


UUMMANNAQ 2016
Nuka-Sofie Fleishcher Lovstrom reclining in front of her son's painting.

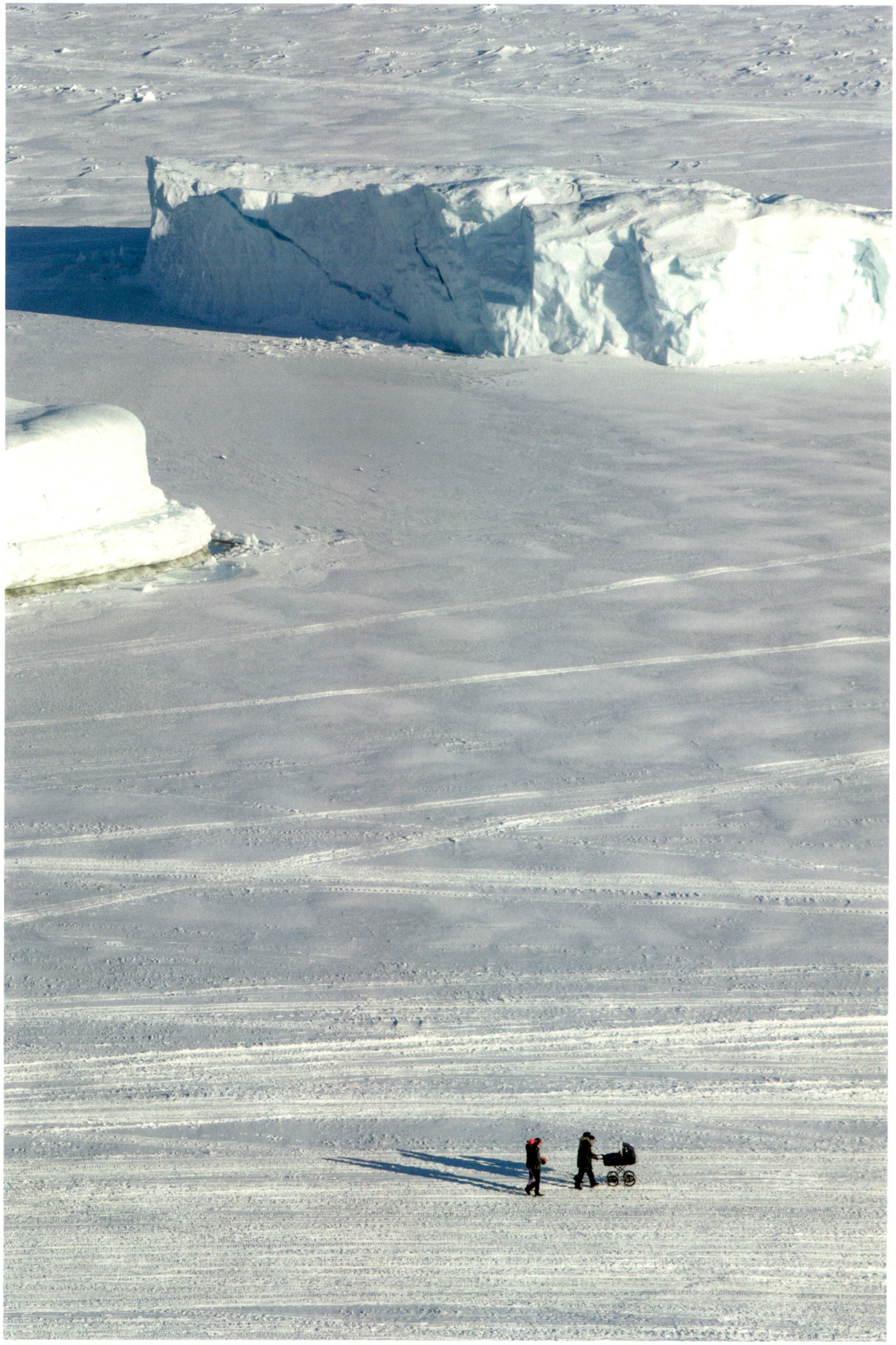

UUMMANNAQ 2017
A Sunday family stroll on the fjord. Sunny days make for a beautiful time to enjoy the promenade.

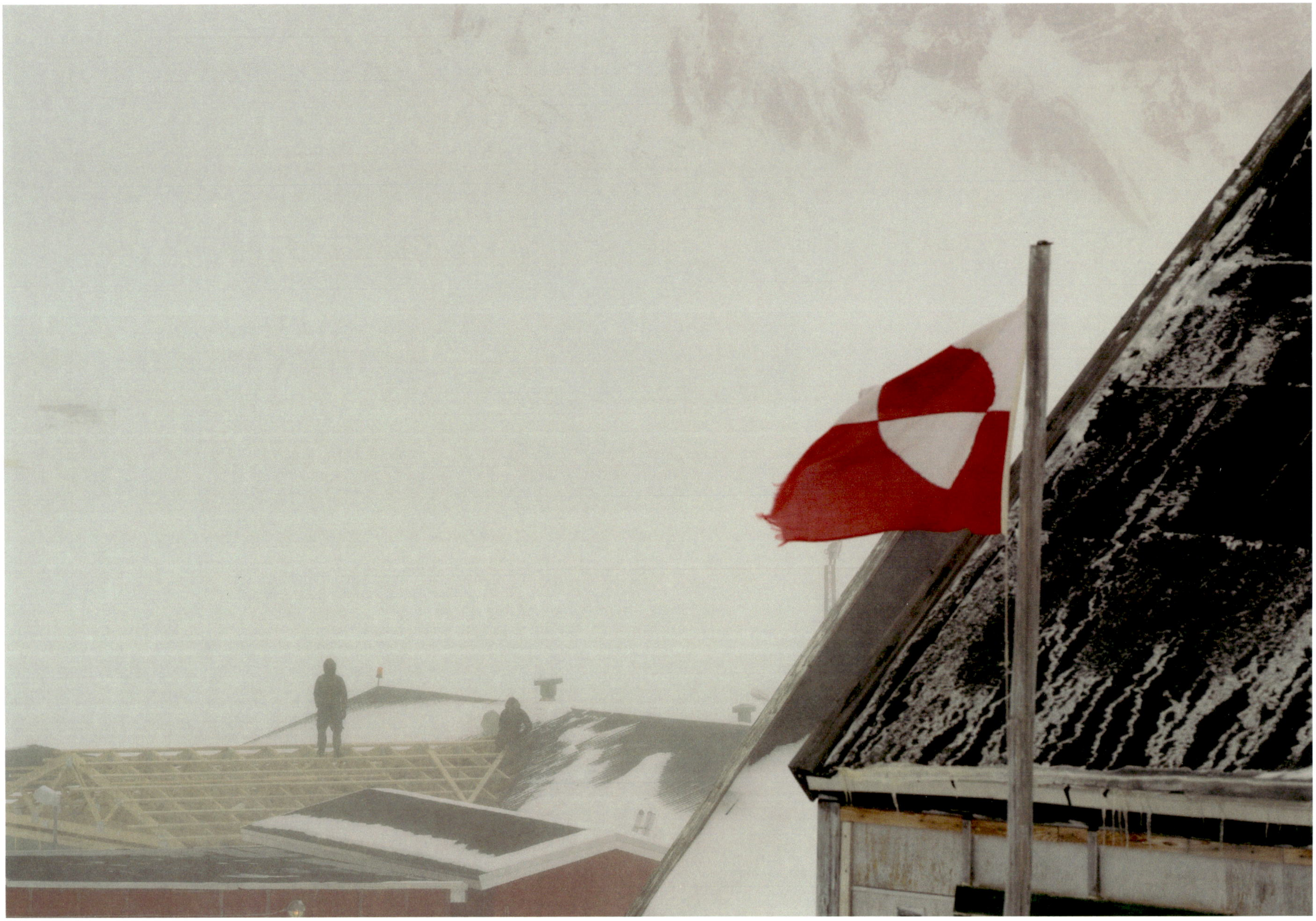

ILLORSUIT 2017
Despite the snowstorm, work continues on the fish factory roof.

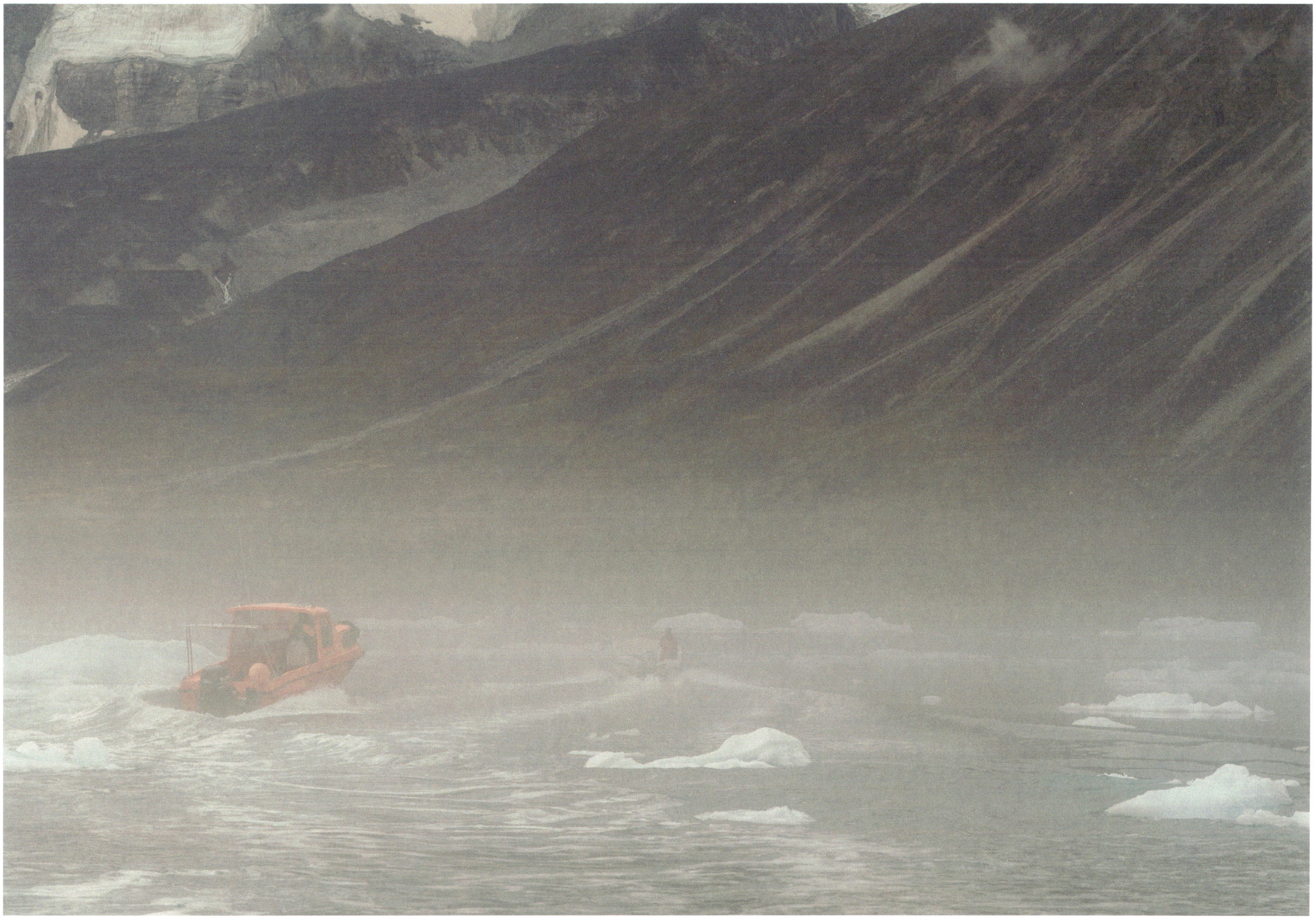

UUMMANNAQ DISTRICT 2016
Sailing the Uummannaq Fjord through ice and fog toward the four-mile-wide Store Glacier.

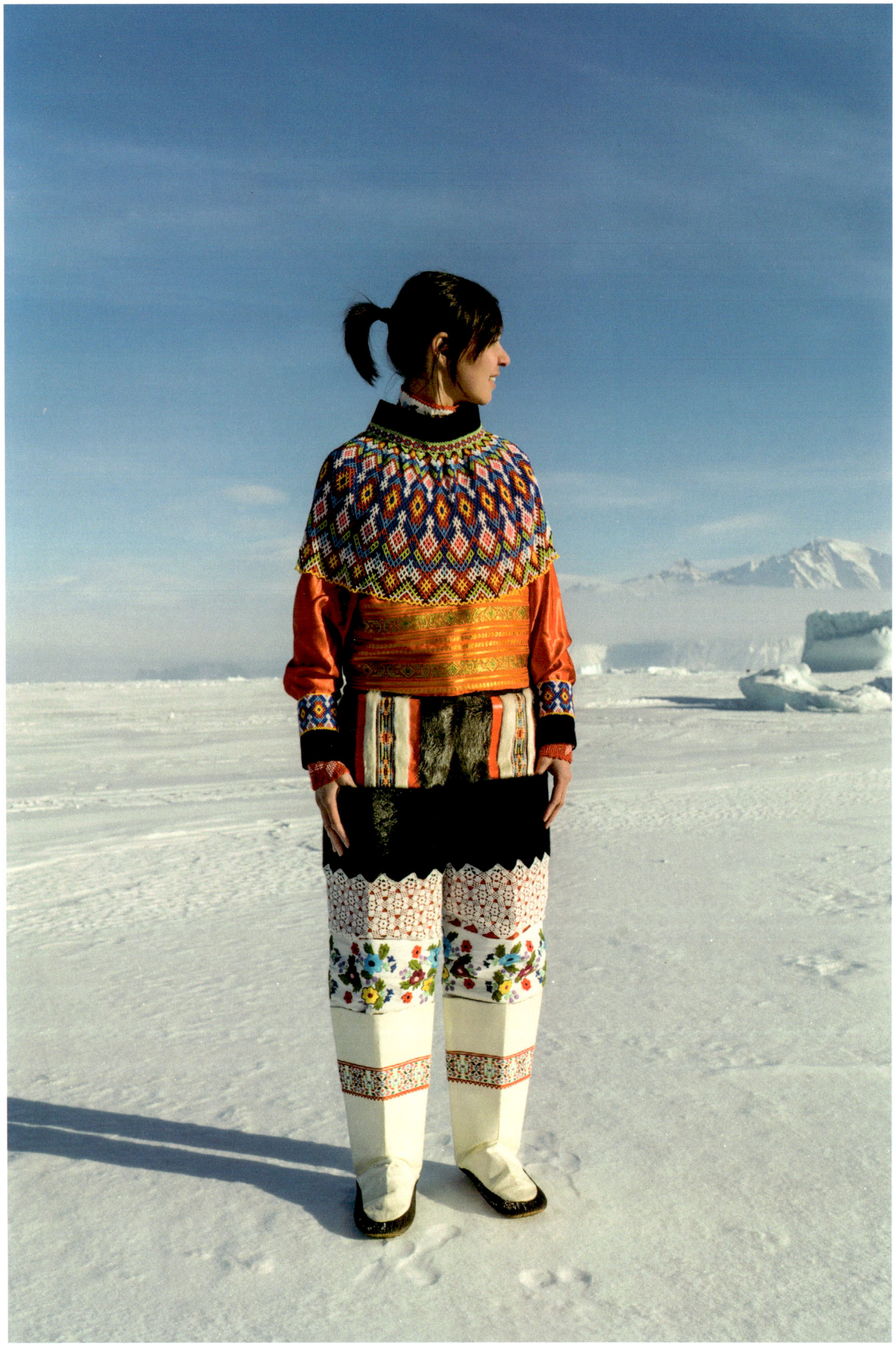

ILLORSUIT 2017

Katrina Zeeb in a traditional Greenlandic outfit worn for special occasions. When Kent lived in Illorsuit, this was the normal attire for women.

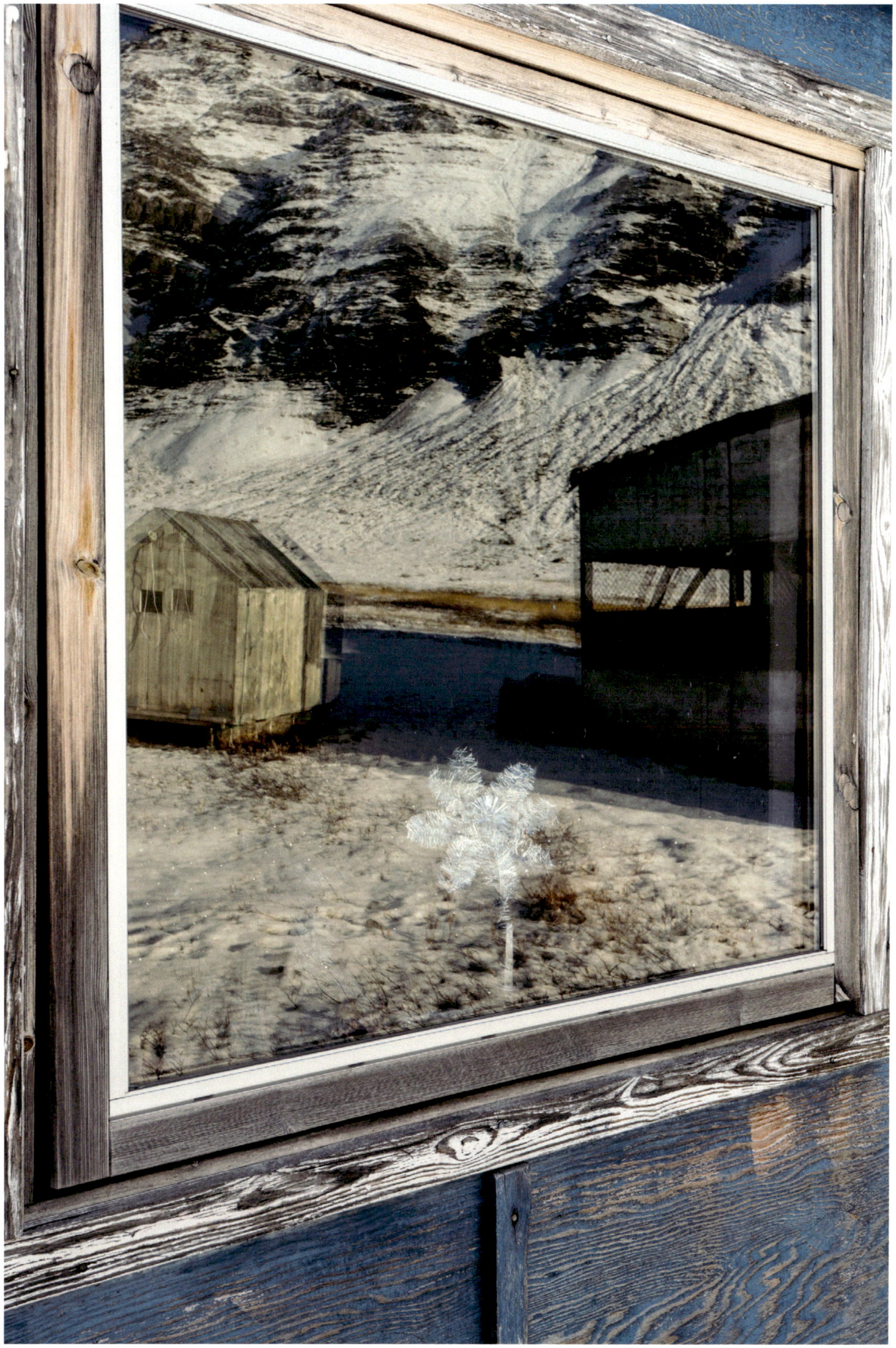

ILLORSUIT 2016

Reflection of mountains west of Illorsuit. Abandoned buildings used to be active sites for processing fish.

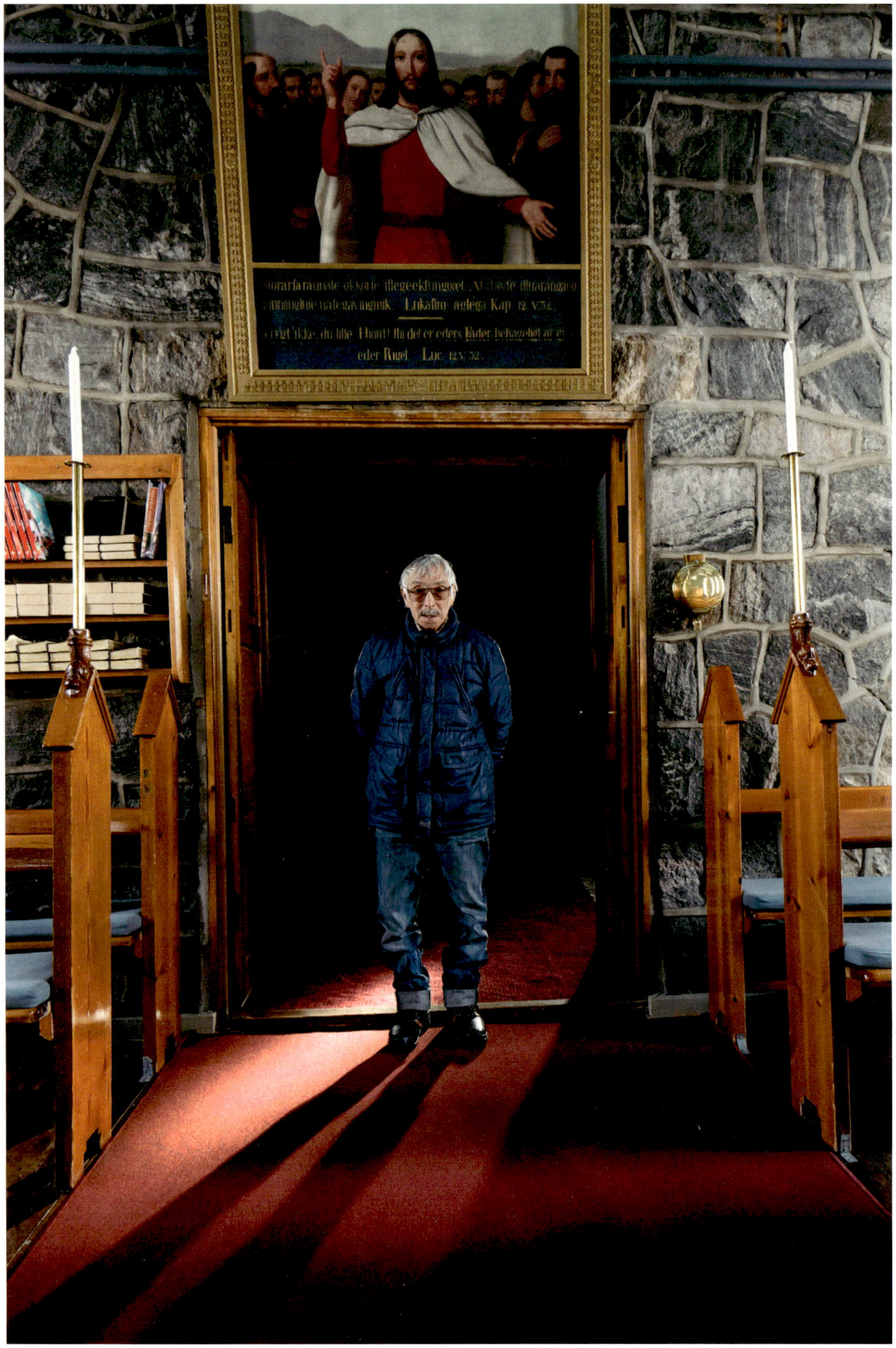


UUMMANNAQ 2017
Custodian George Orstrom in the Uummannaq church.

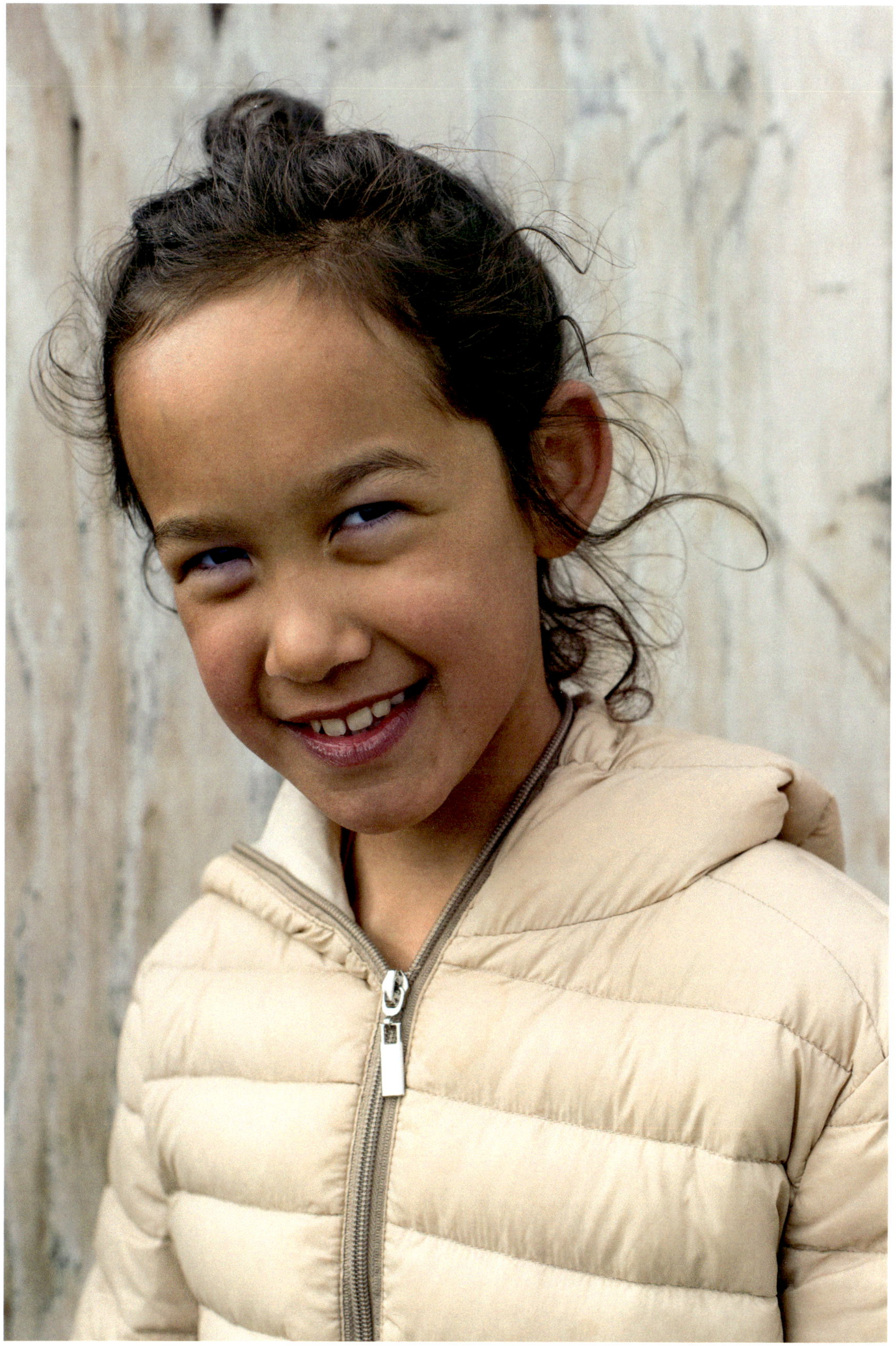

ILLORSUIT 2016
Majka Qvist

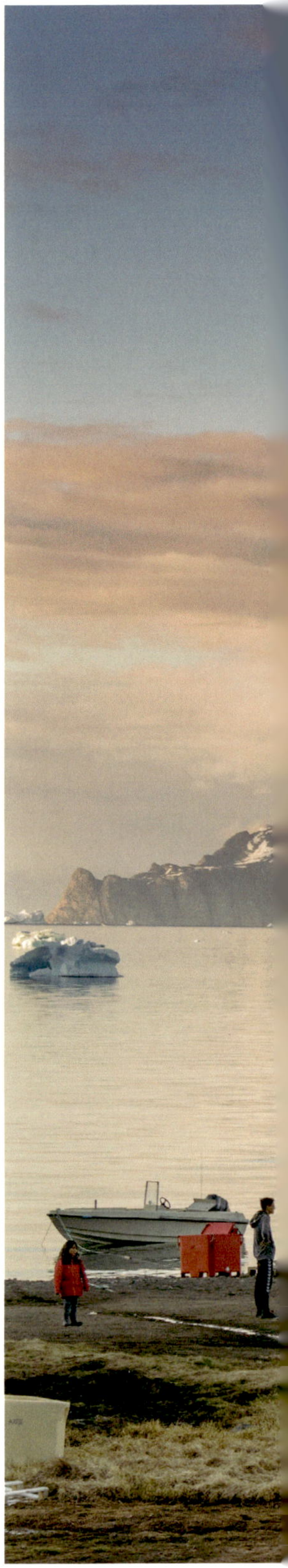

ILLORSUIT 2016
Each spring when the weather warms up, young and old alike play the baseball-like game of rounders.

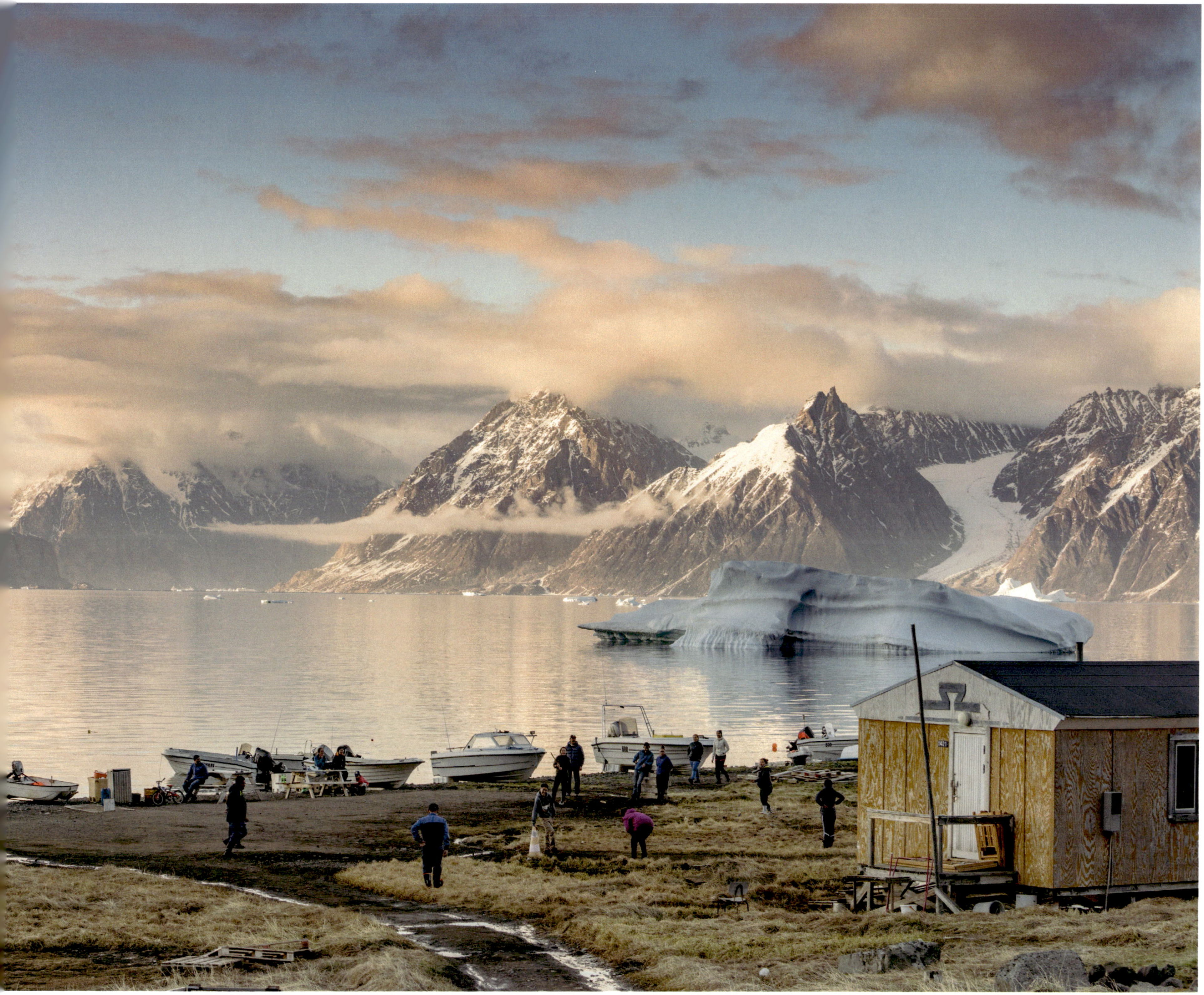

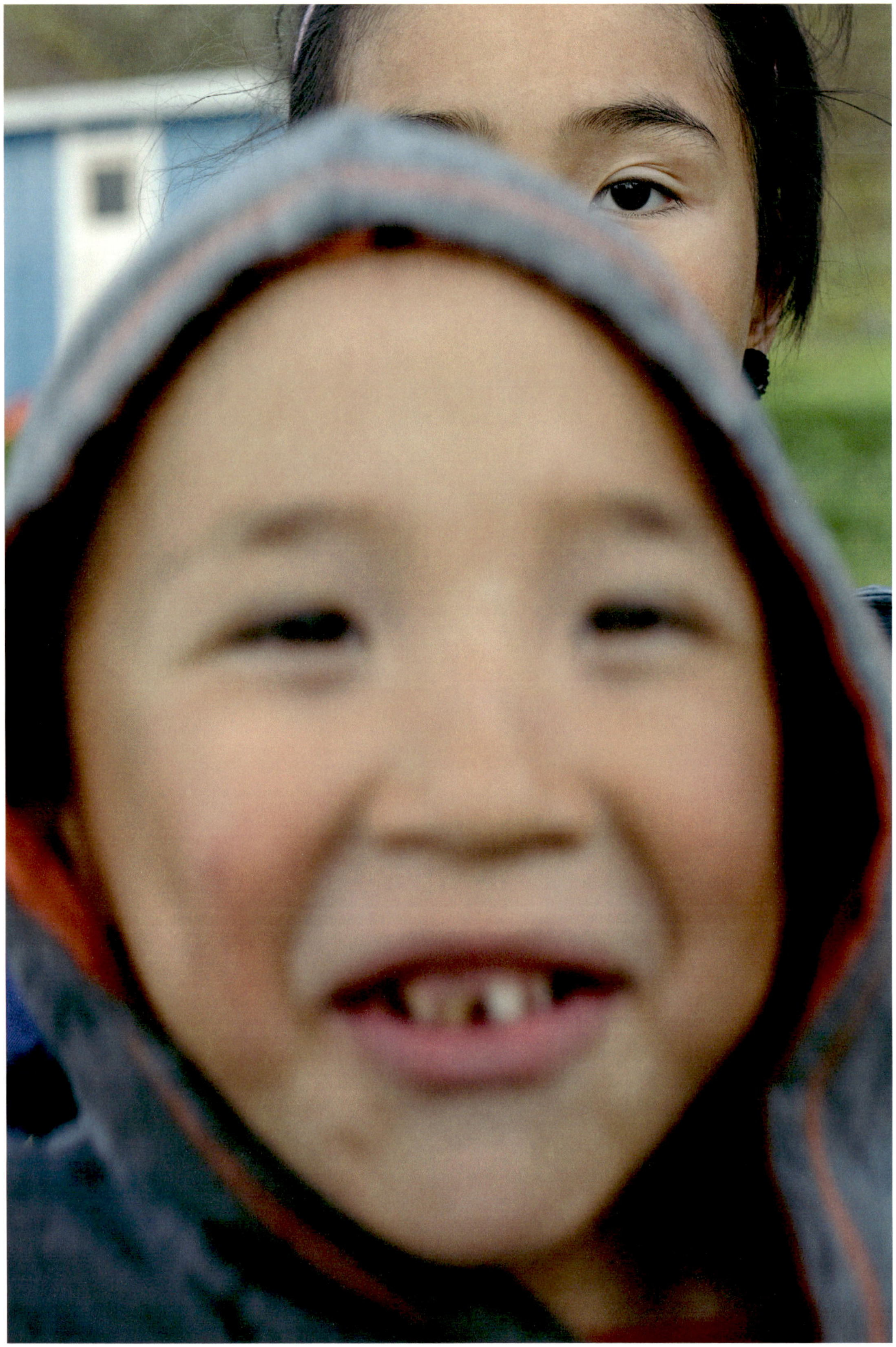

ILLORSUIT 2016
Frederik and Najannguak Malakiassen smile for the camera.

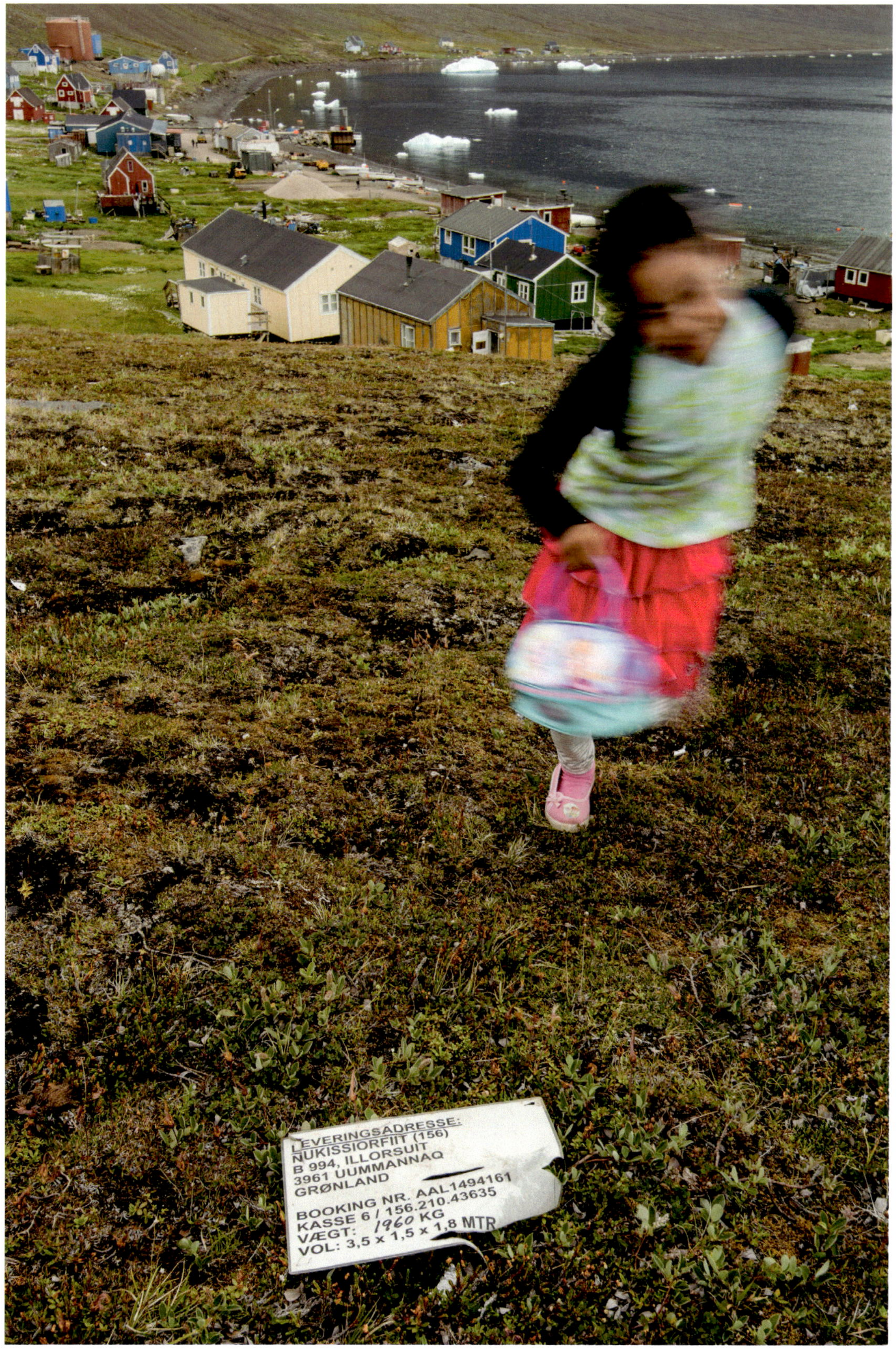


ILLORSUIT 2016

A government plaque for Illorsuit with the settlement and bay in the background.

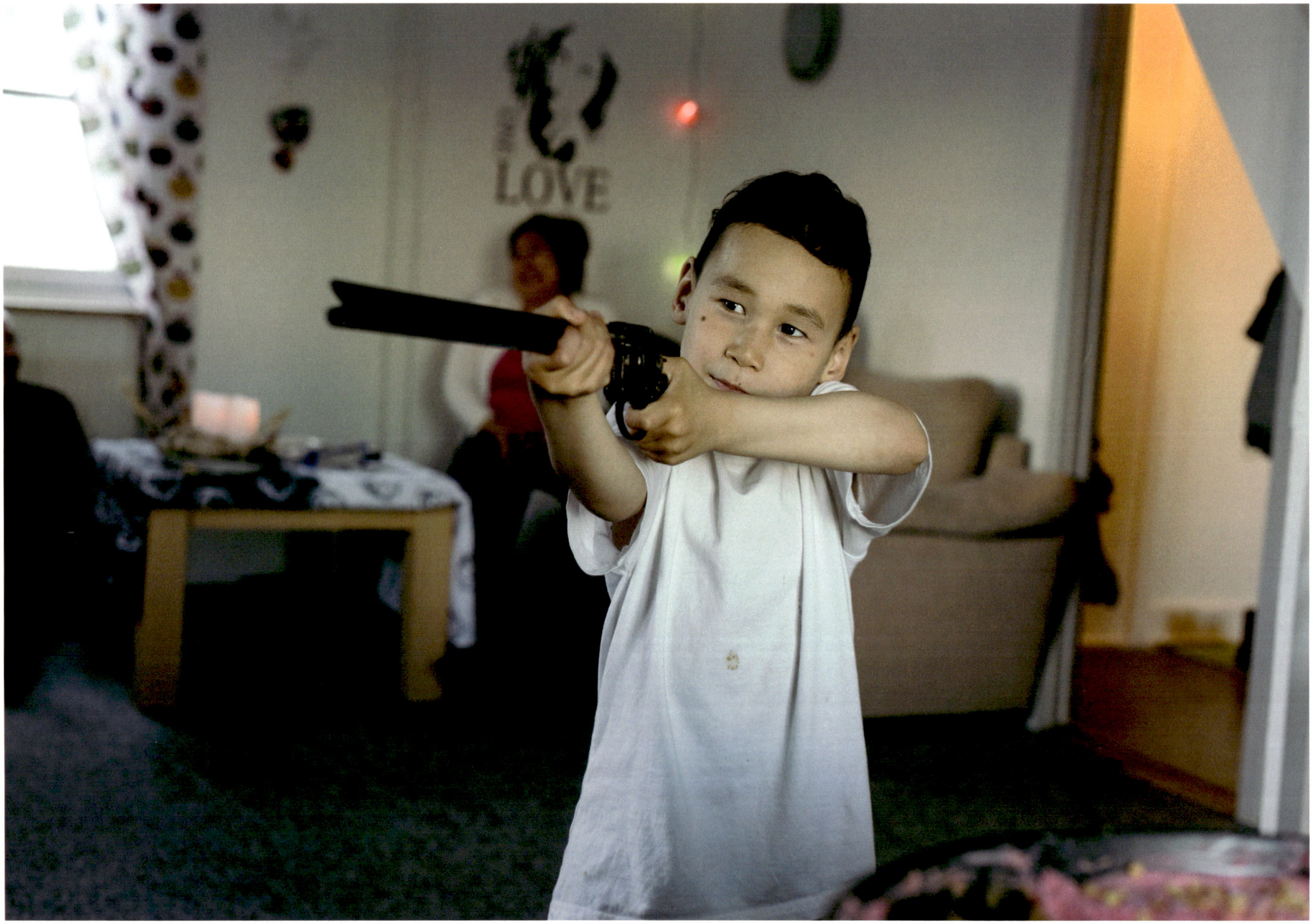


ILLORSUIT 2016
*Kaffemik* birthday party for Uiloriaq M. Jeremiassen, who plays with a toy gun as visitors chat.

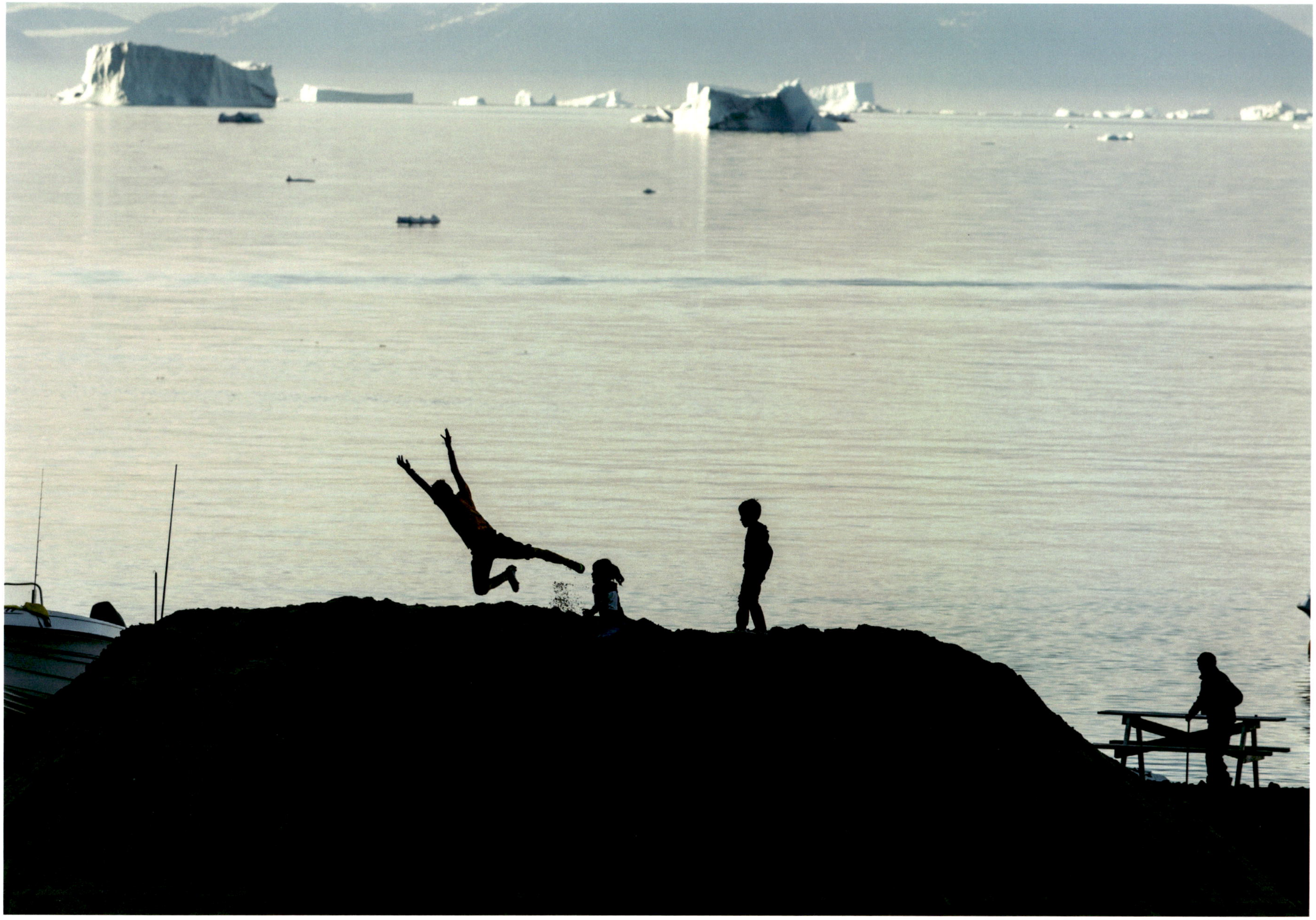

ILLORSUIT 2016

A pile of sand becomes a temporary playground for children.

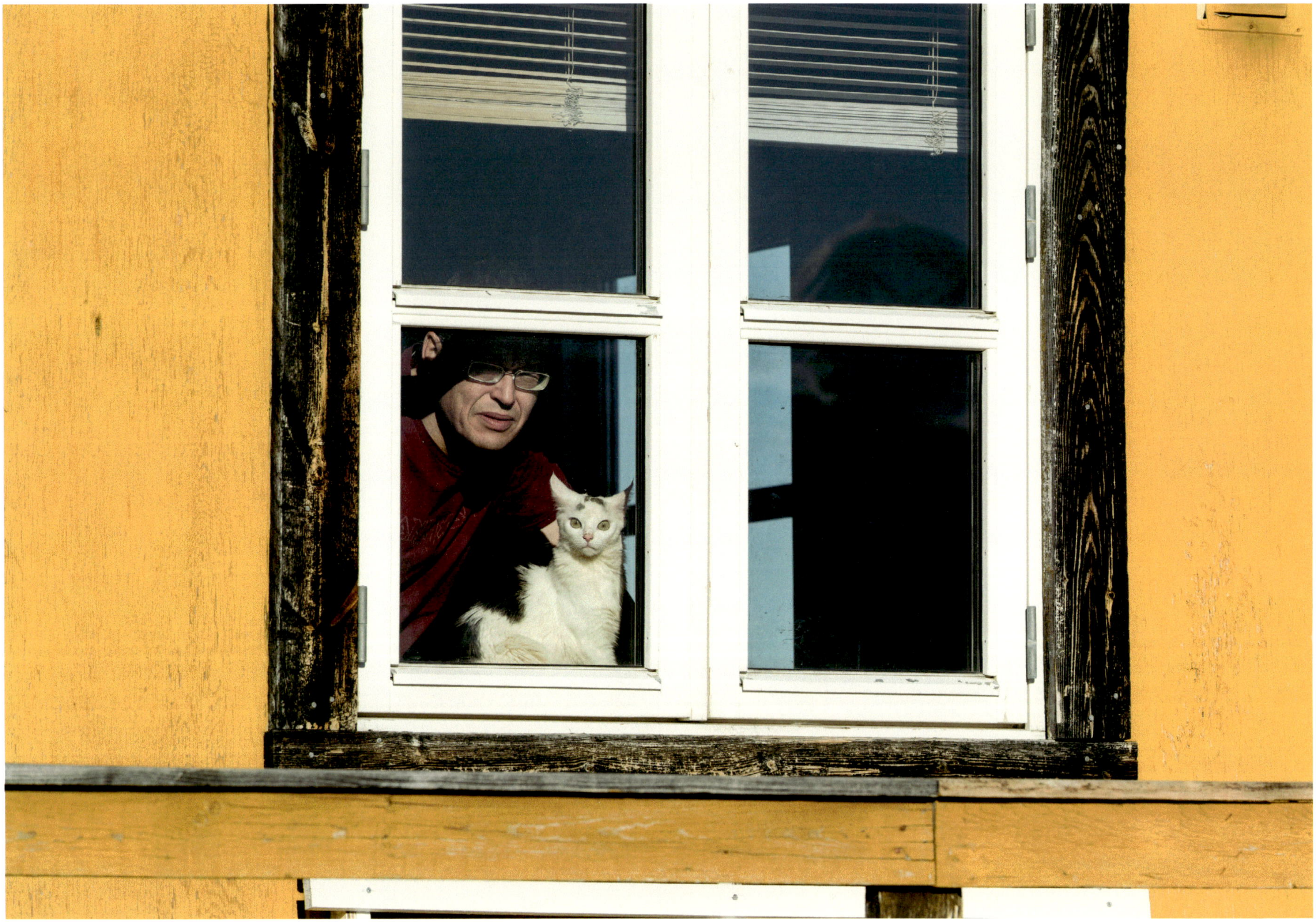

UUMMANNAQ 2016
Mads Fleischer looks out the window with his cat. Few people in Greenland have cats, and most are indoor pets.

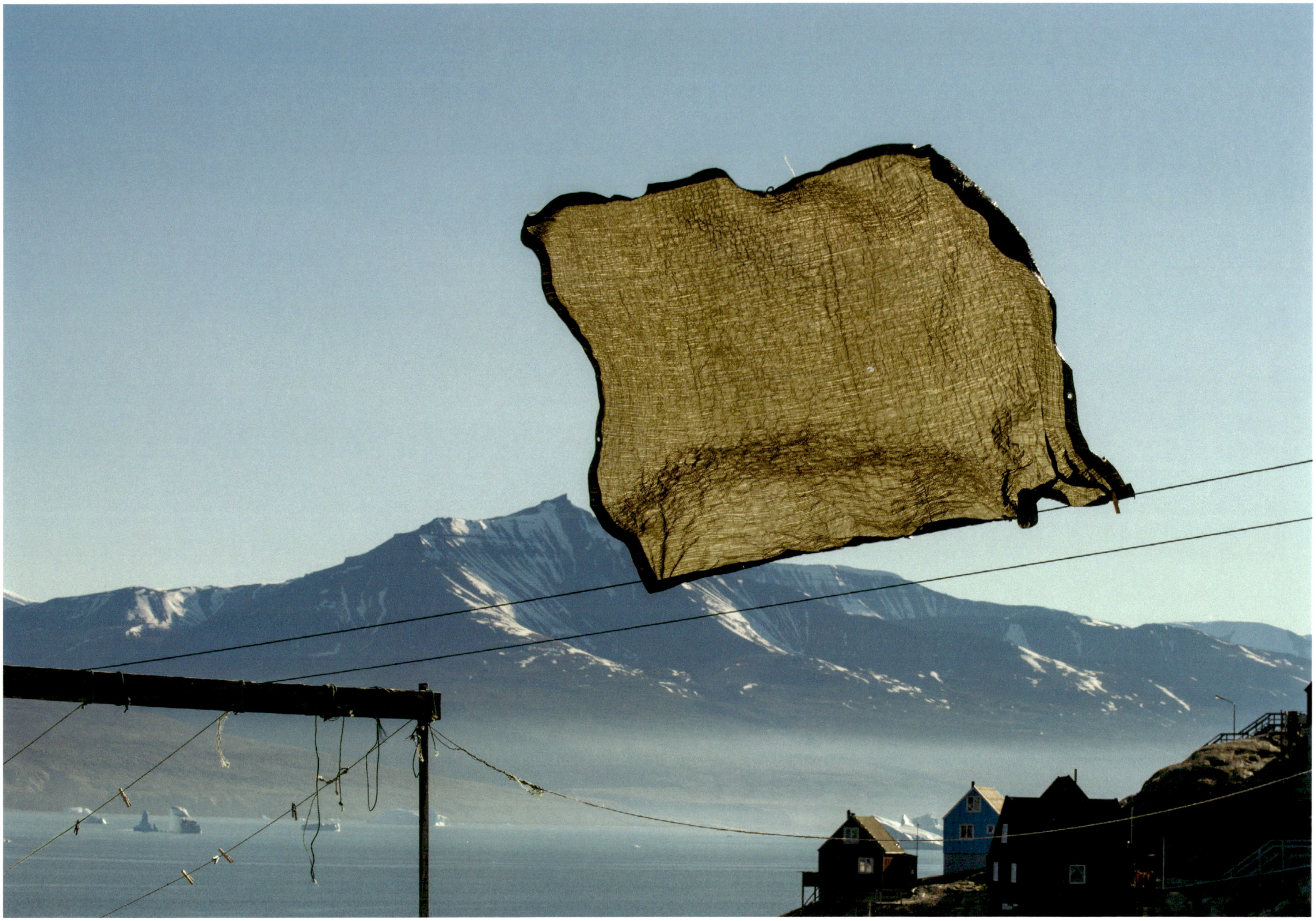

UUMMANNAQ 2016
An airborne tarp contrasts against the mountains and blue summer sky.

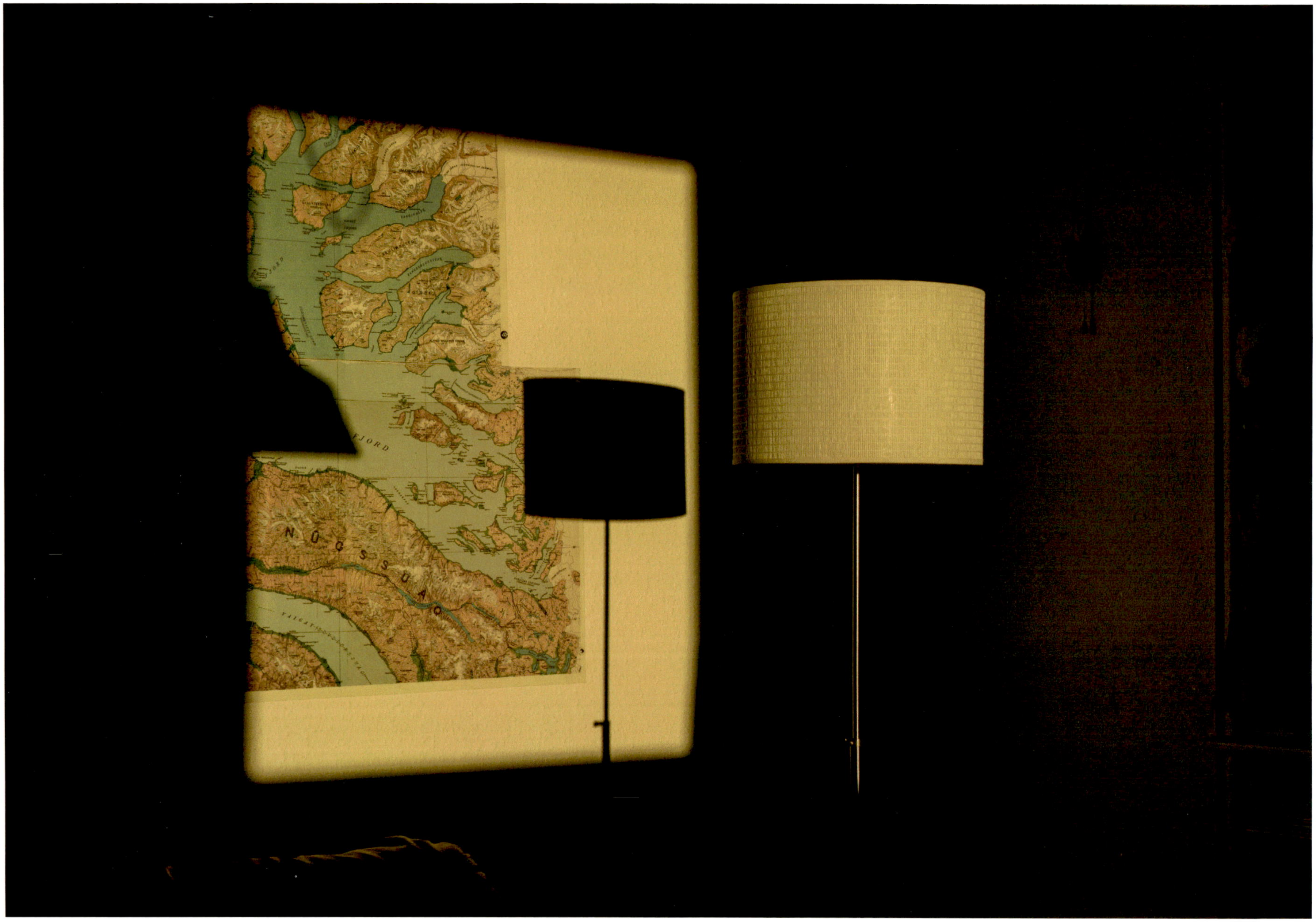


UUMMANNAQ 2017

Warm winter sun strikes an Uummannaq District map and lamp in a hillside apartment.

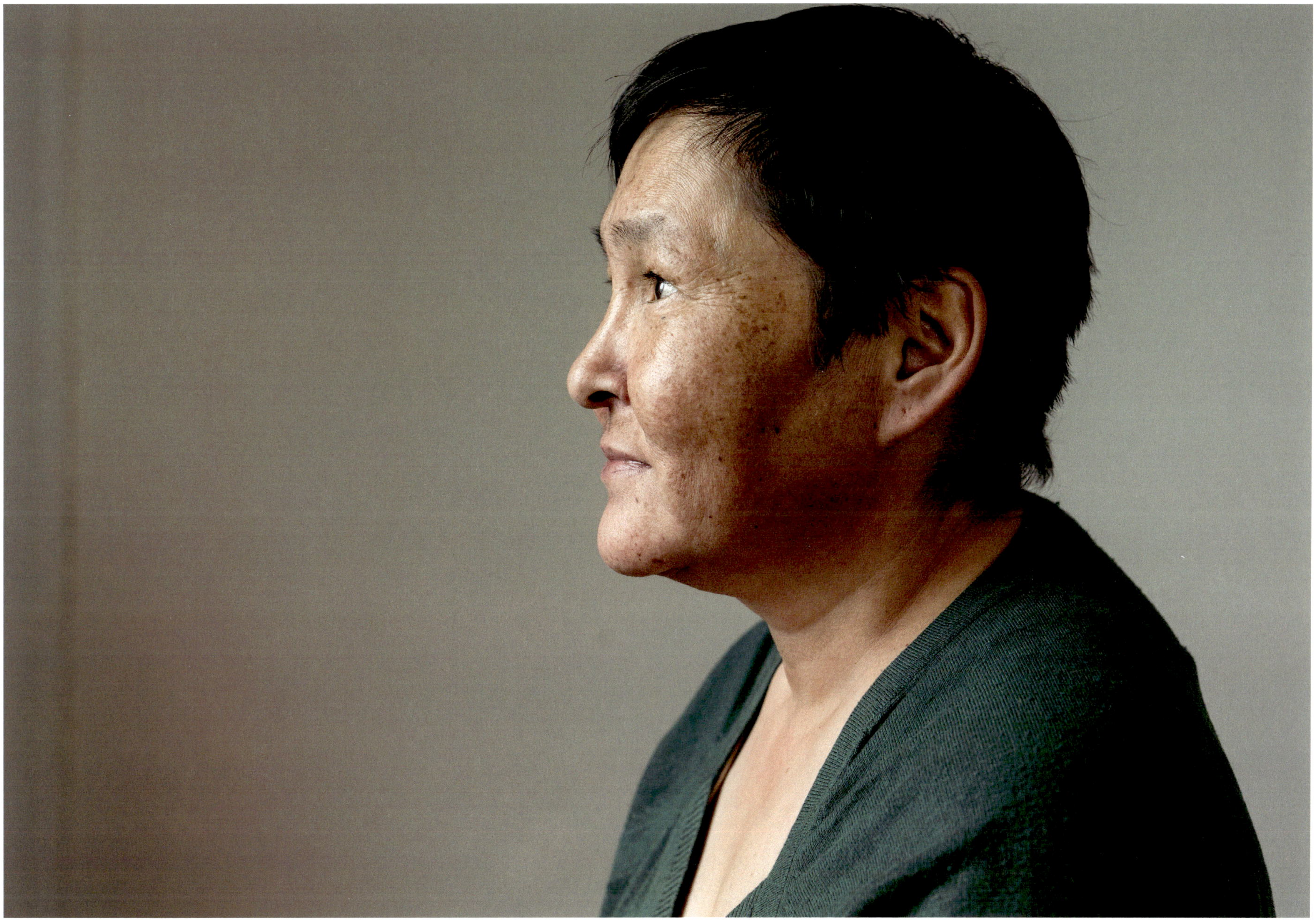

ILLORSUIT 2017
Lea Korneliussen, one of many Korneliussens in Illorsuit. She is a manager at the local *Pilersuisoq* grocery store.

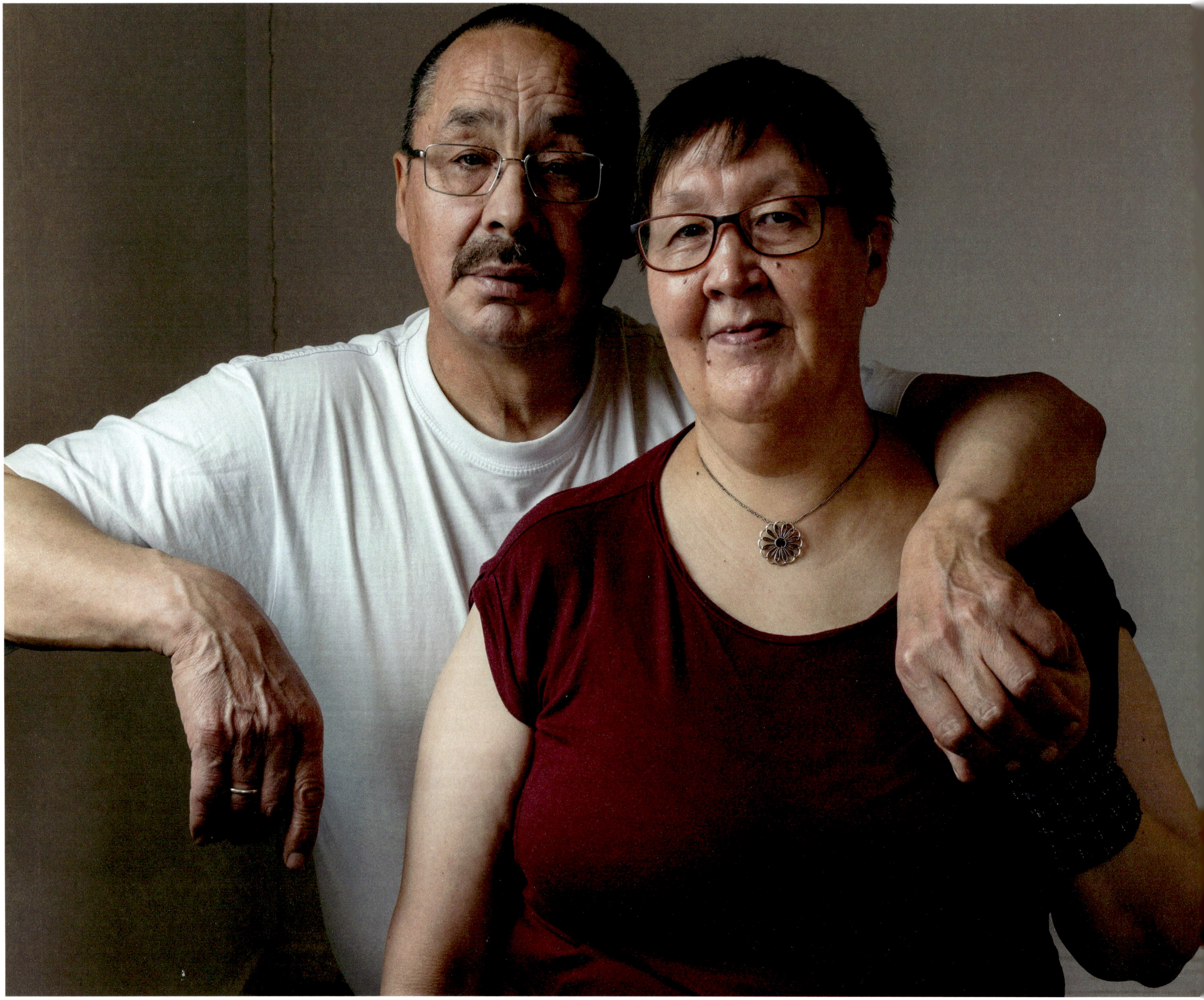

ILLORSUIT 2017

Illorsuit Settlement Managers Jarte and Kasper Korneliussen take care of the office, gasoline pumps and heating fuel, helipad, and community activities.

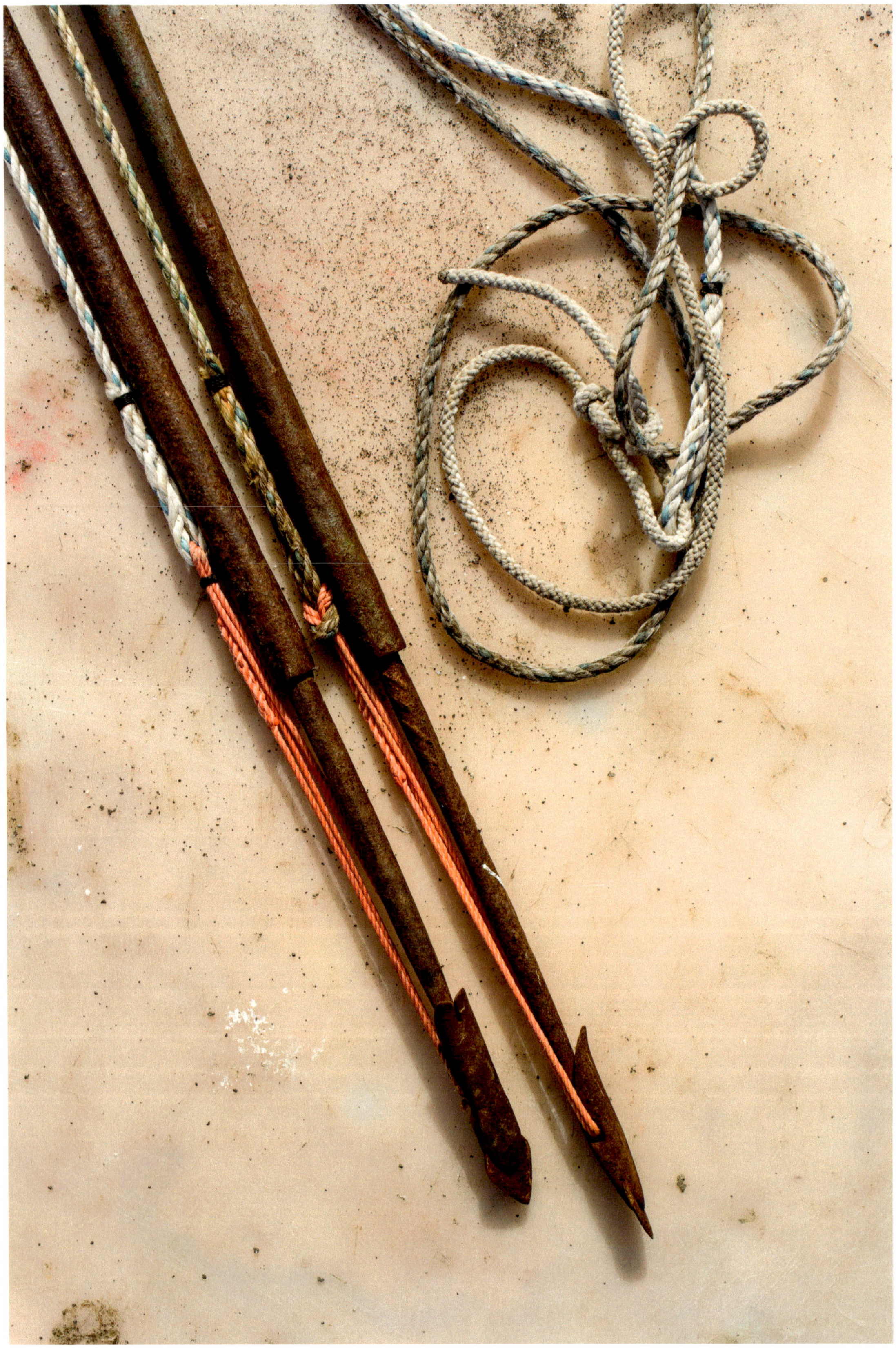

ILLORSUIT 2016
Harpoons and rope are now seldom used for fishing but are available if needed.

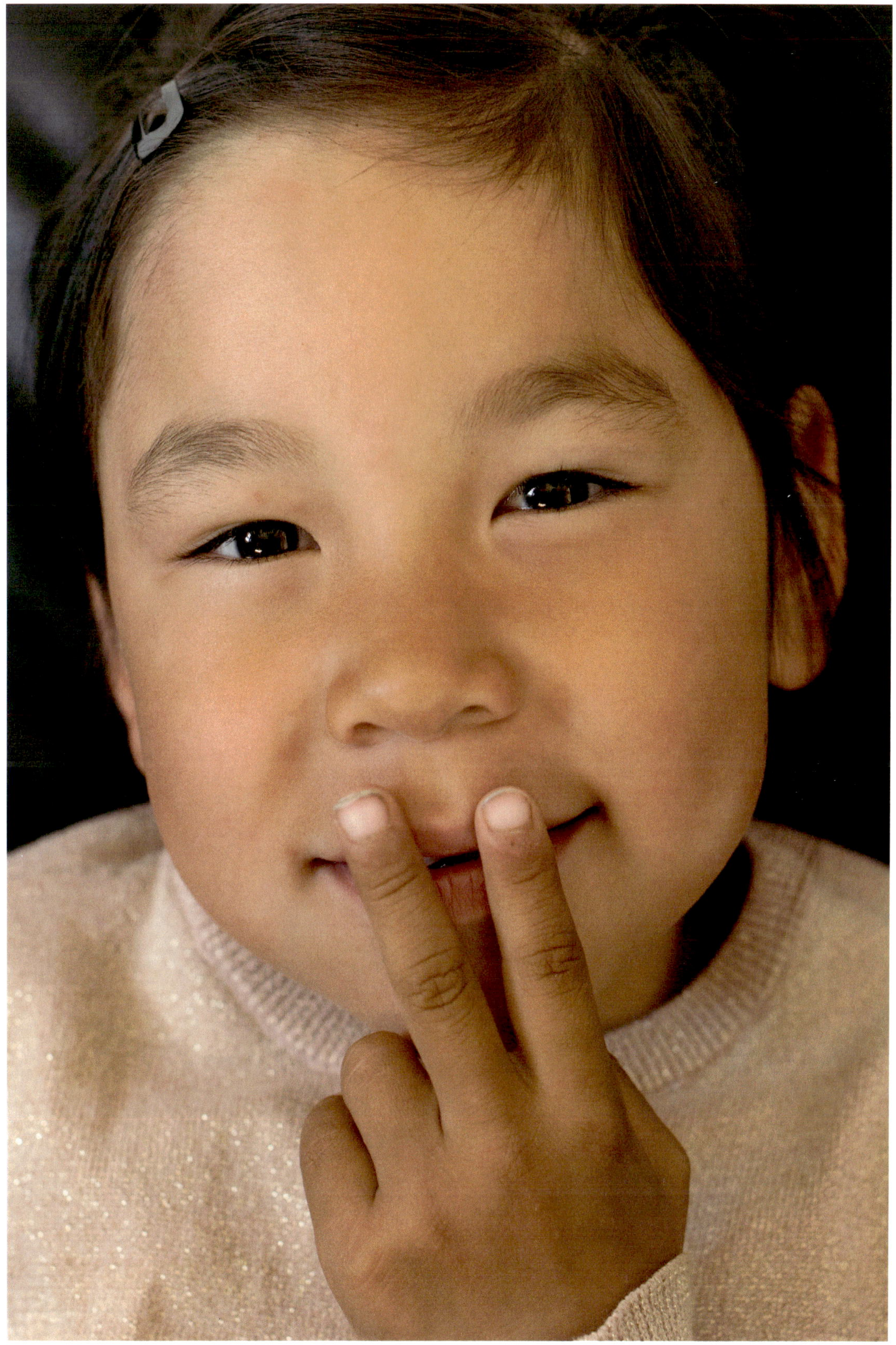

ILLORSUIT 2016
Francis K. Jonassen, named after Frances Kent, Rockwell's wife.

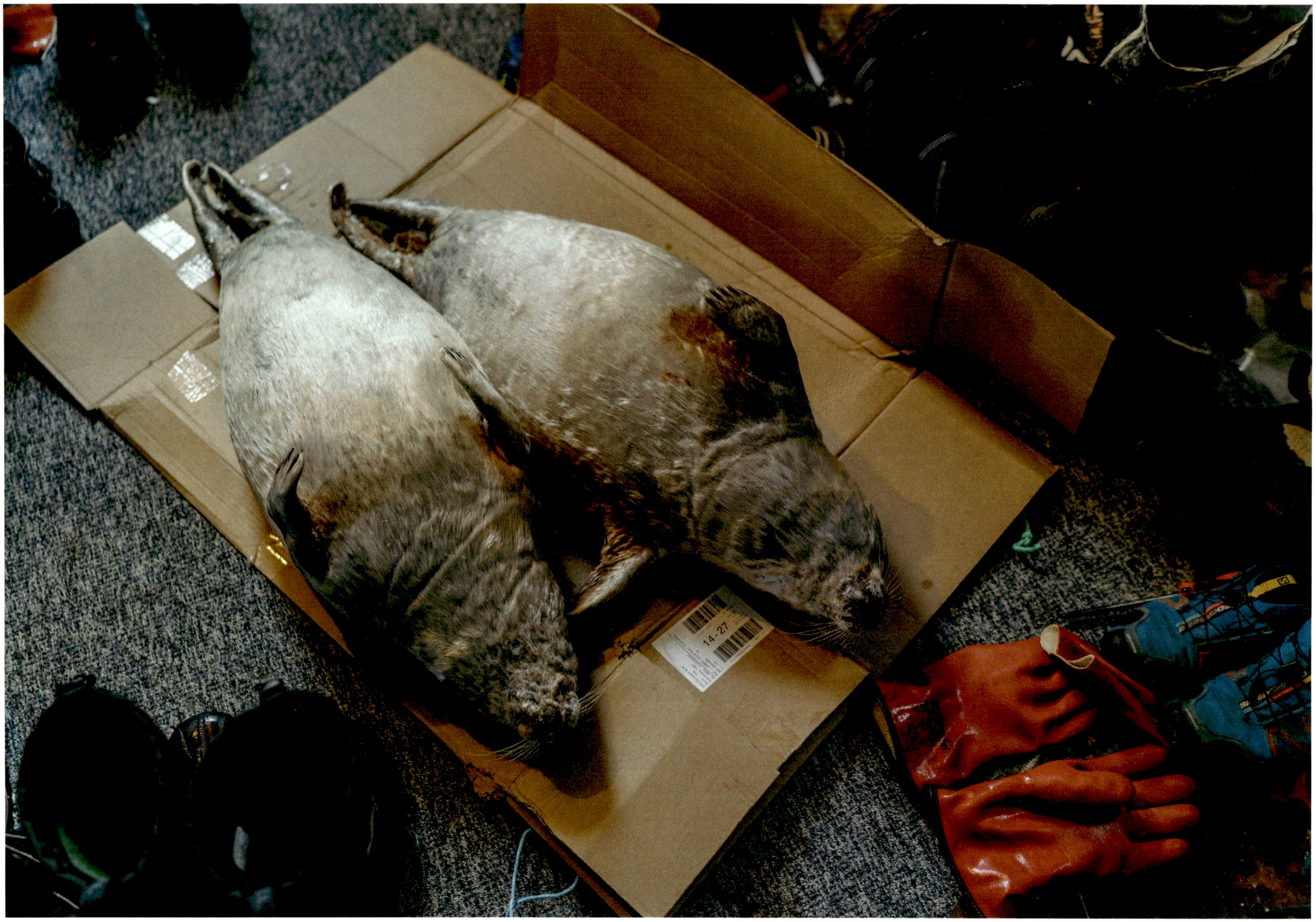

ILLORSUIT 2017
Seals thawing in the Korneliussen coatroom. Once thawed, they will be skinned and fed to dogs.

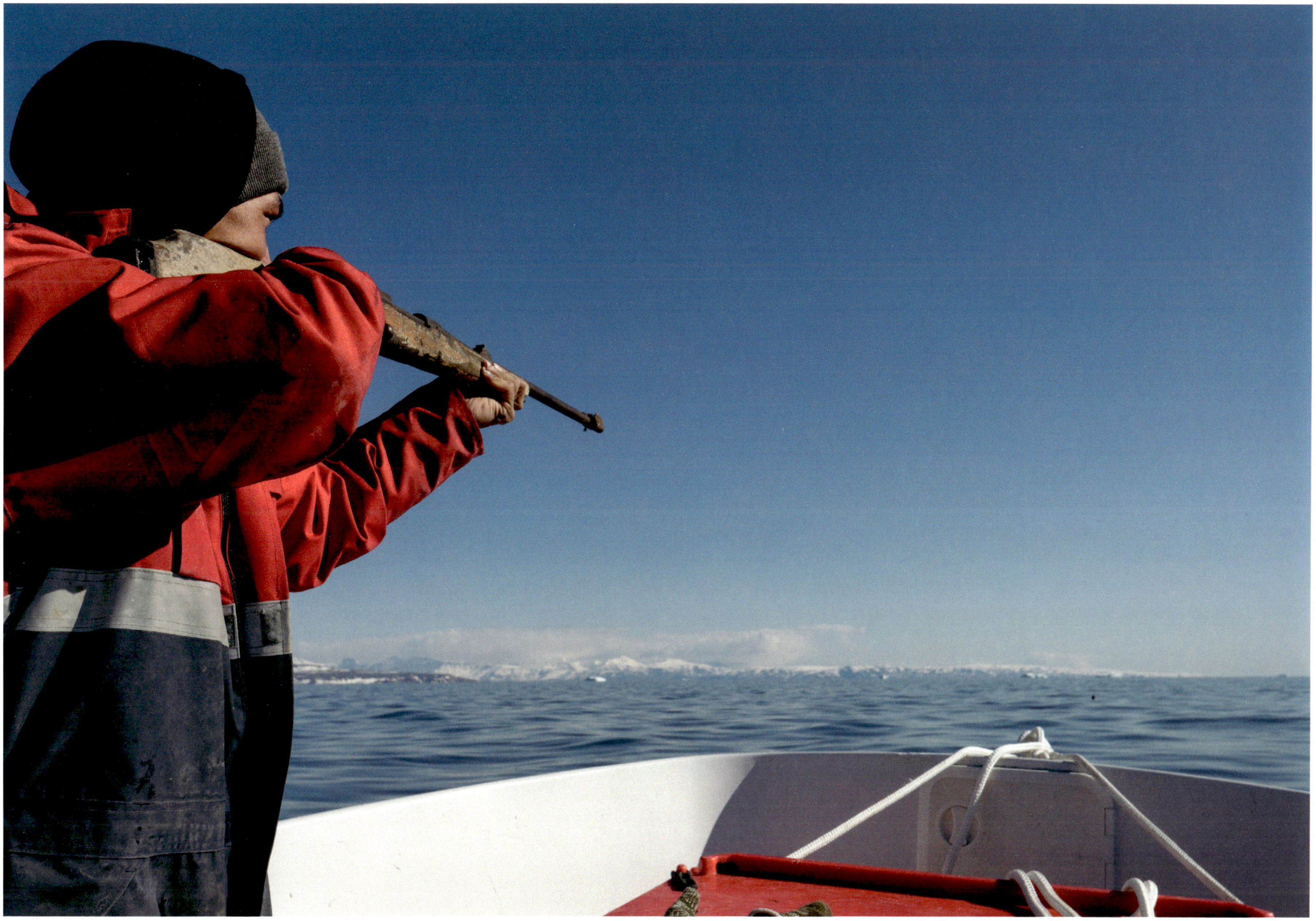

ILLORSUIT SOUND 2016
Peter Jonassen taking aim at a seal. The small black speck is hard to hit, with the boat bouncing and the seal above water for only a few seconds.

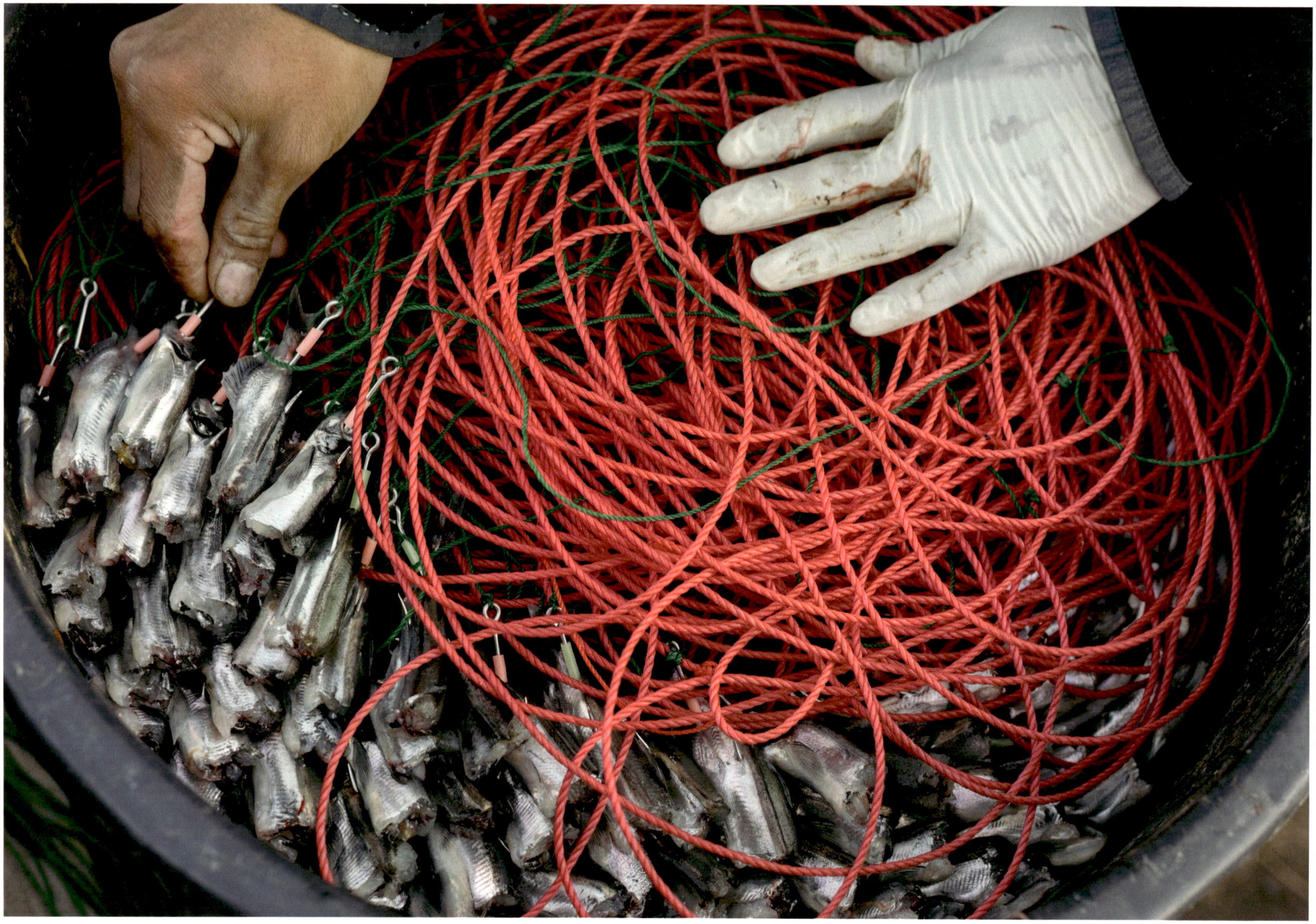

ILLORSUIT 2016

Baiting long-line hooks with capelin fish. The line may be more than 1,000 yards long with as many as 1,100 hooks.

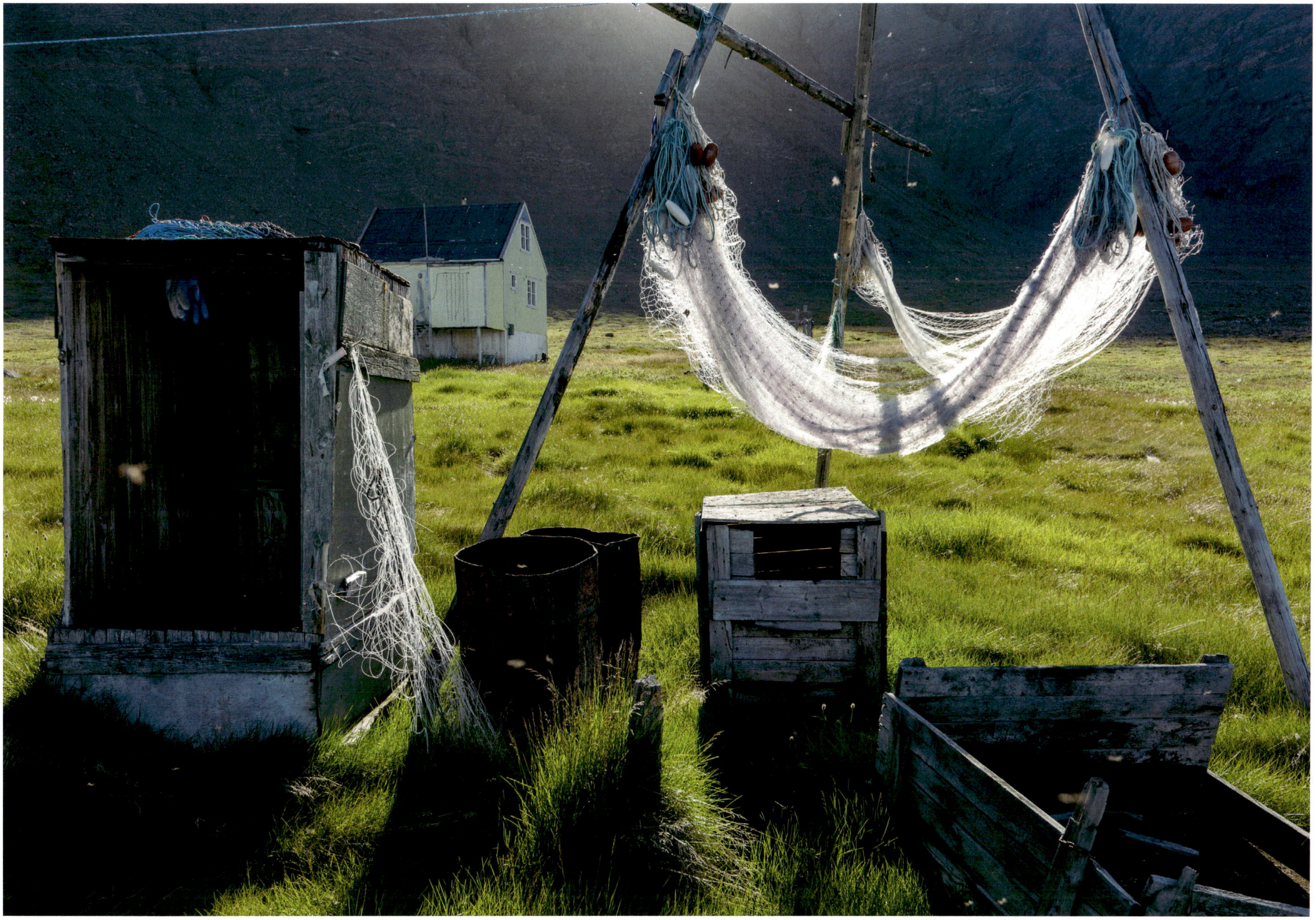

ILLORSUIT 2016
A fishing net dries in the sunset.

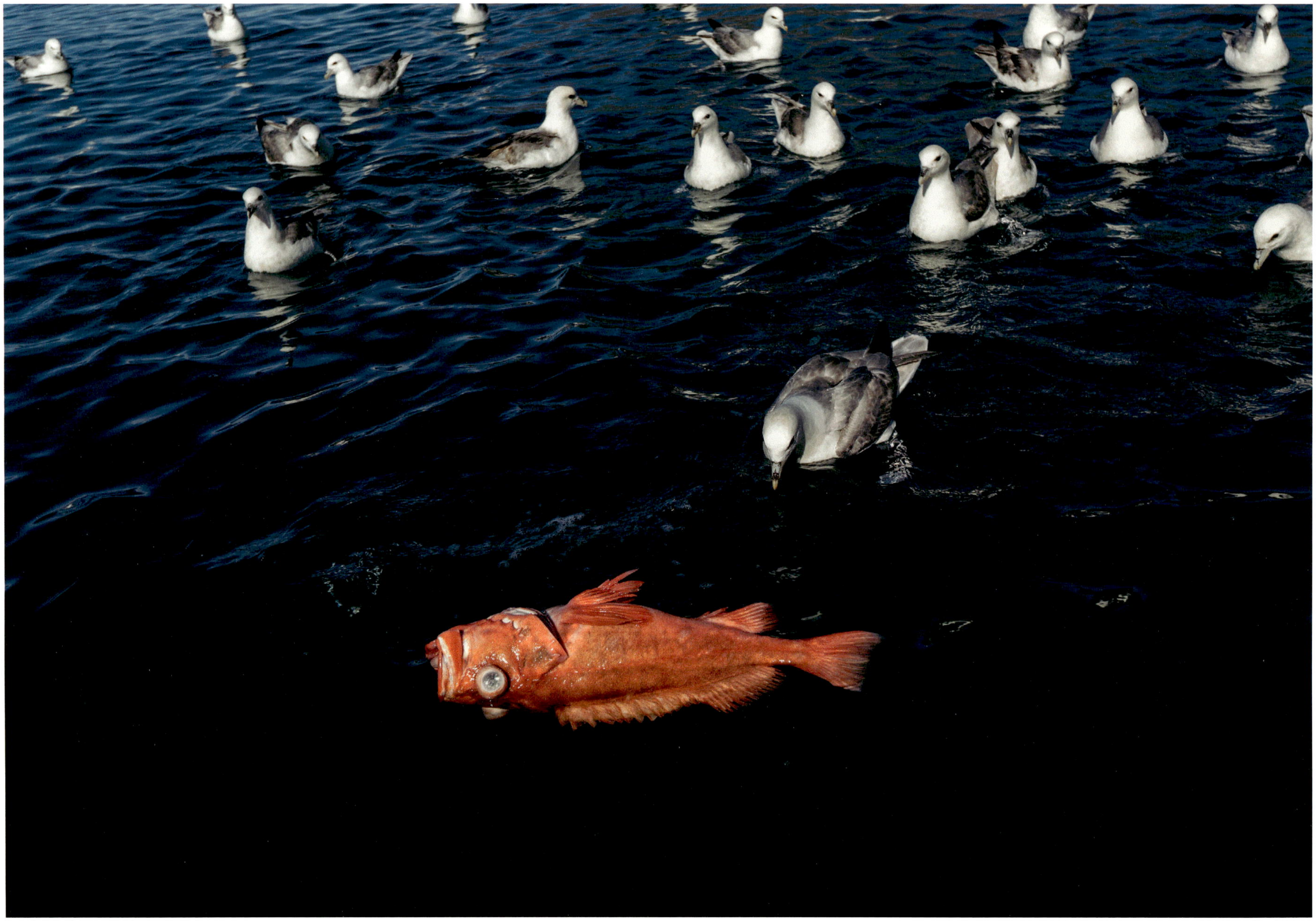

ILLORSUIT SOUND 2016
A redfish is returned to the fjord as a snack for the seagulls.

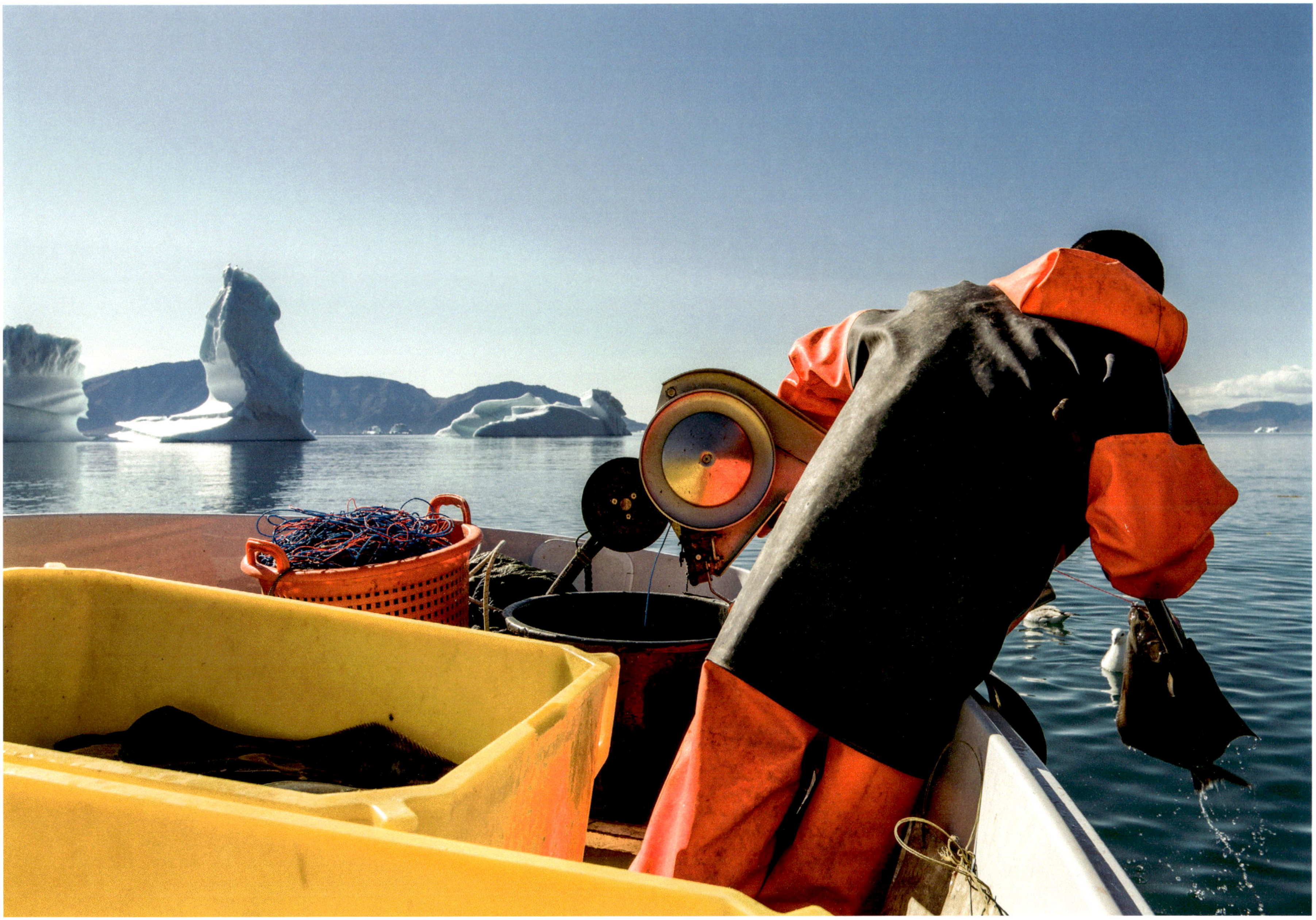

ILLORSUIT SOUND 2016
Mathias Nielsen on an eight-hour day of fishing for halibut. He caught more than 2,000 pounds of fish.

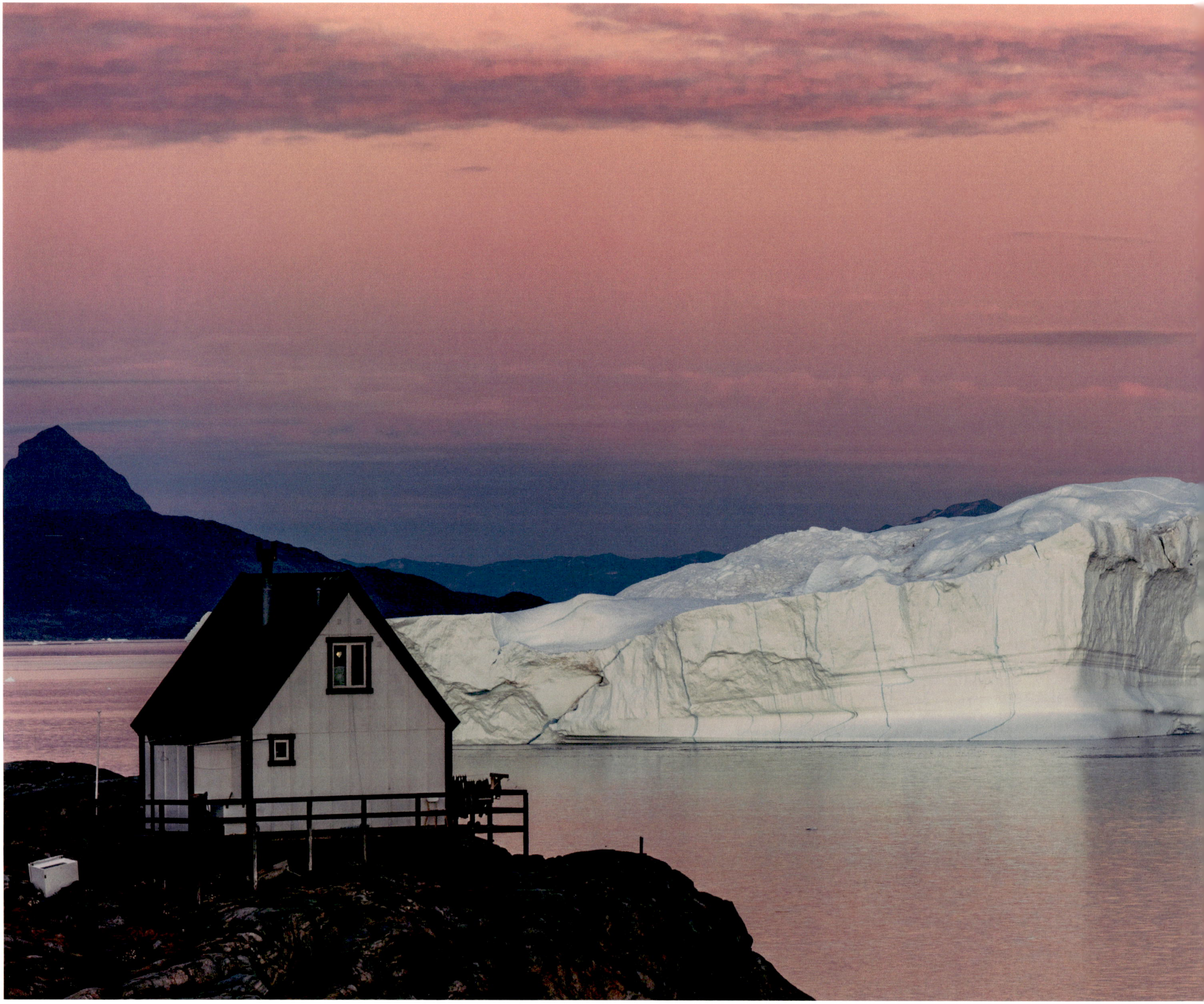

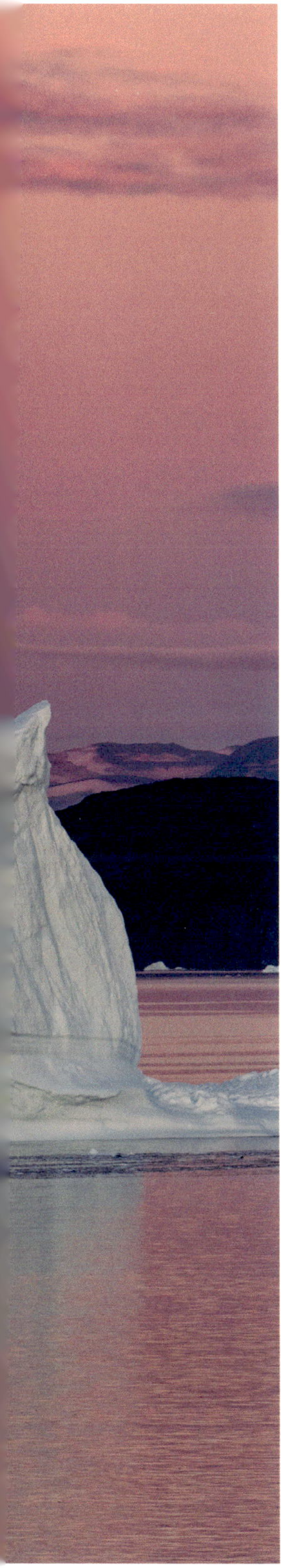

UUMMANNAQ FJORD 2016
The magenta sunset colors of the Sound evoke a watercolor painting.

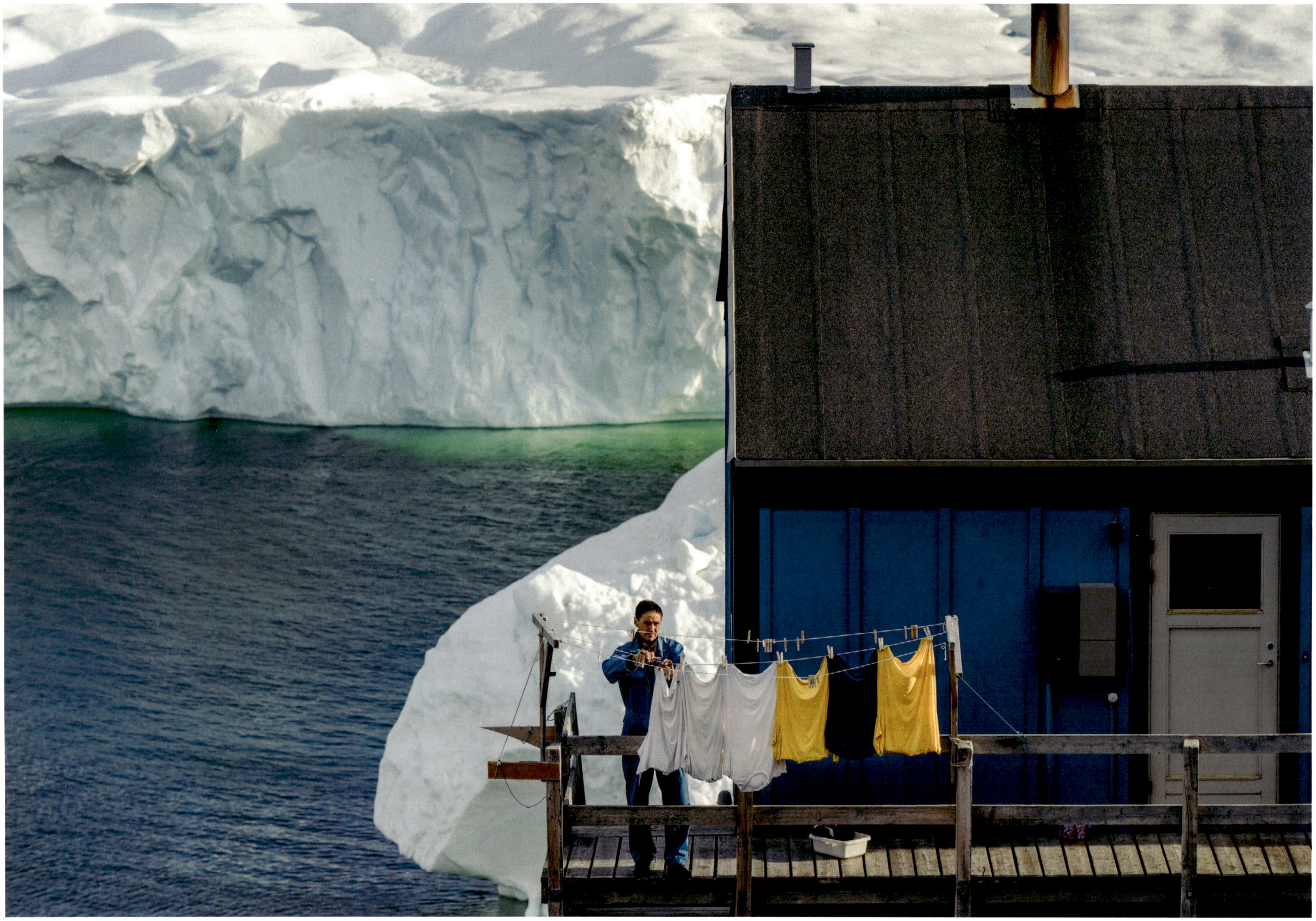

UUMMANNAQ 2016

Clothes are air dried even in the middle of winter—Greenland homes do not have dryers.

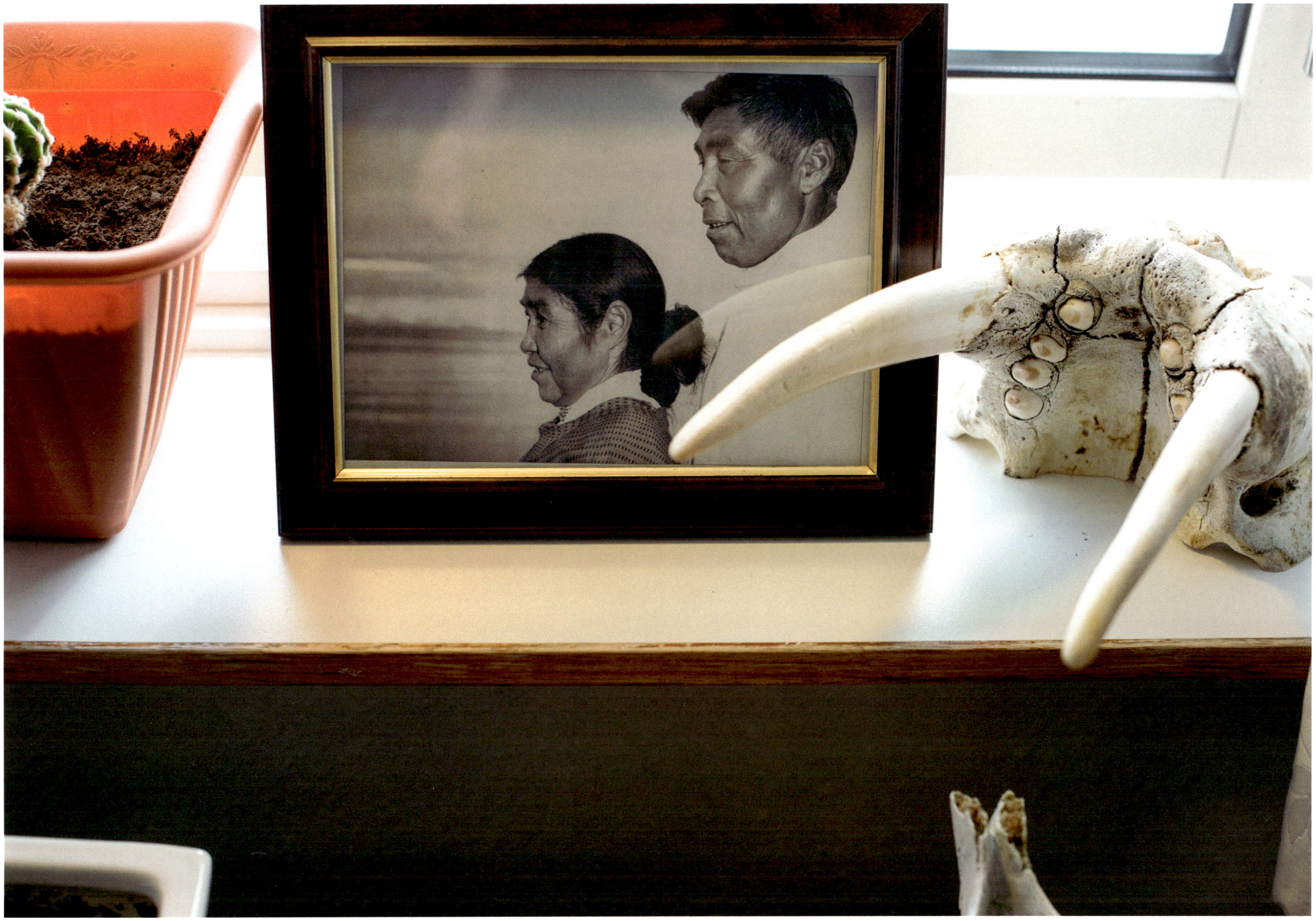

UUMMANNAQ 2016

Family photographs are important to Inuit culture. Photographs of special occasions and family portraits often are on display in homes.

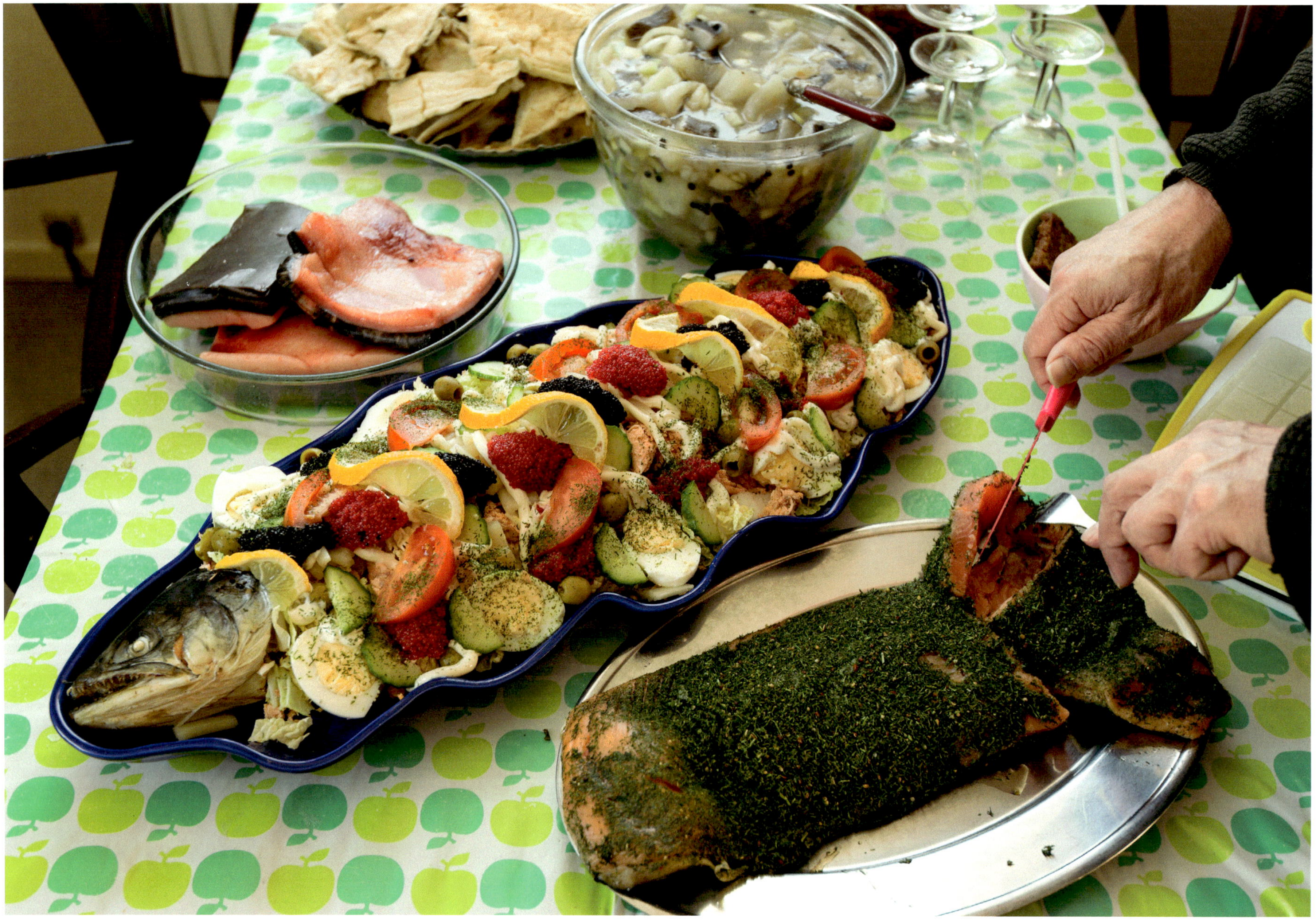

UUMMANNAQ 2016
*Kaffemik* celebrations are a great time to feast on traditional Greenlandic foods.

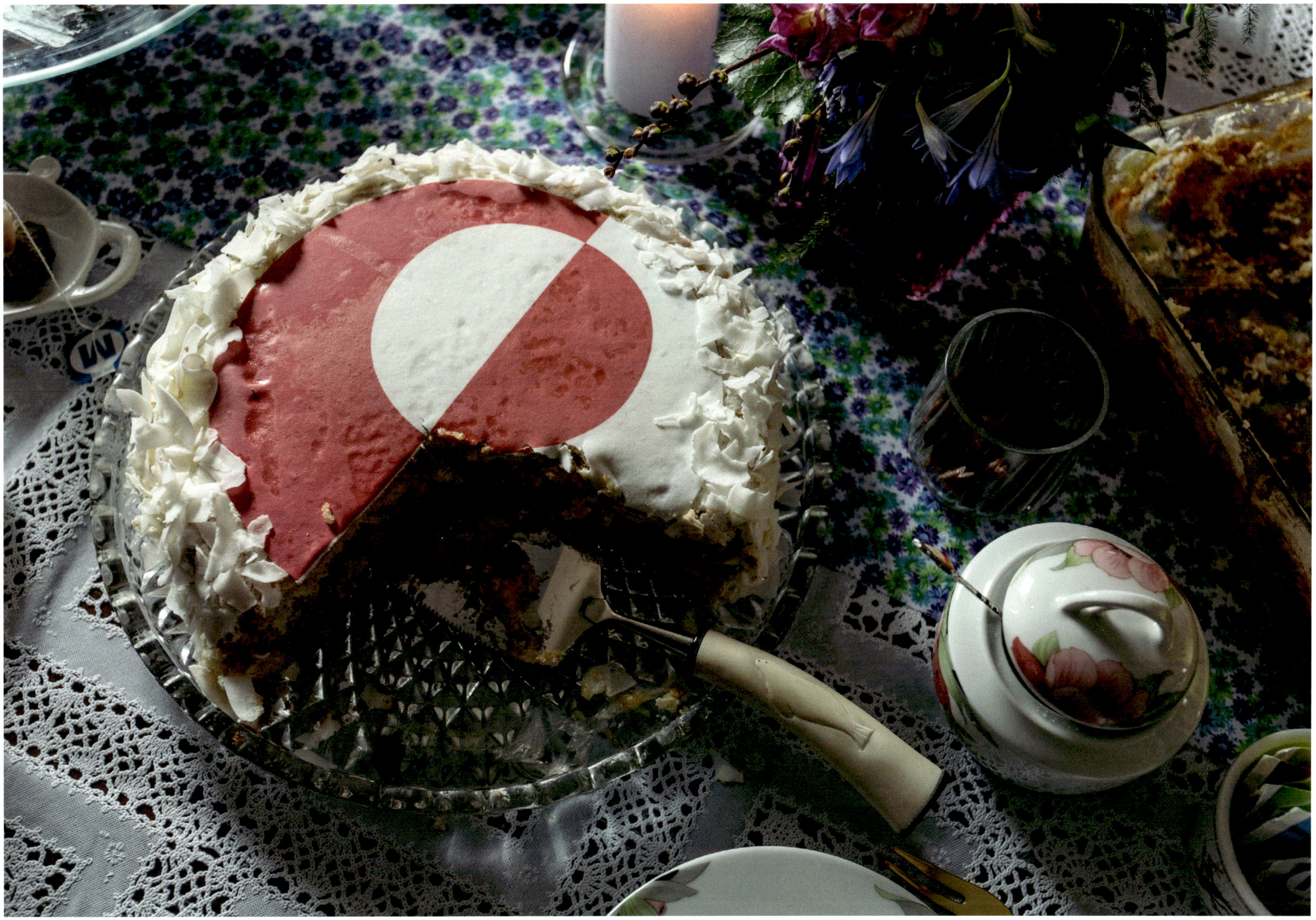

UUMMANNAQ 2016

Greenland's national flag is featured on a *Kaffemik* cake that disappeared quickly.

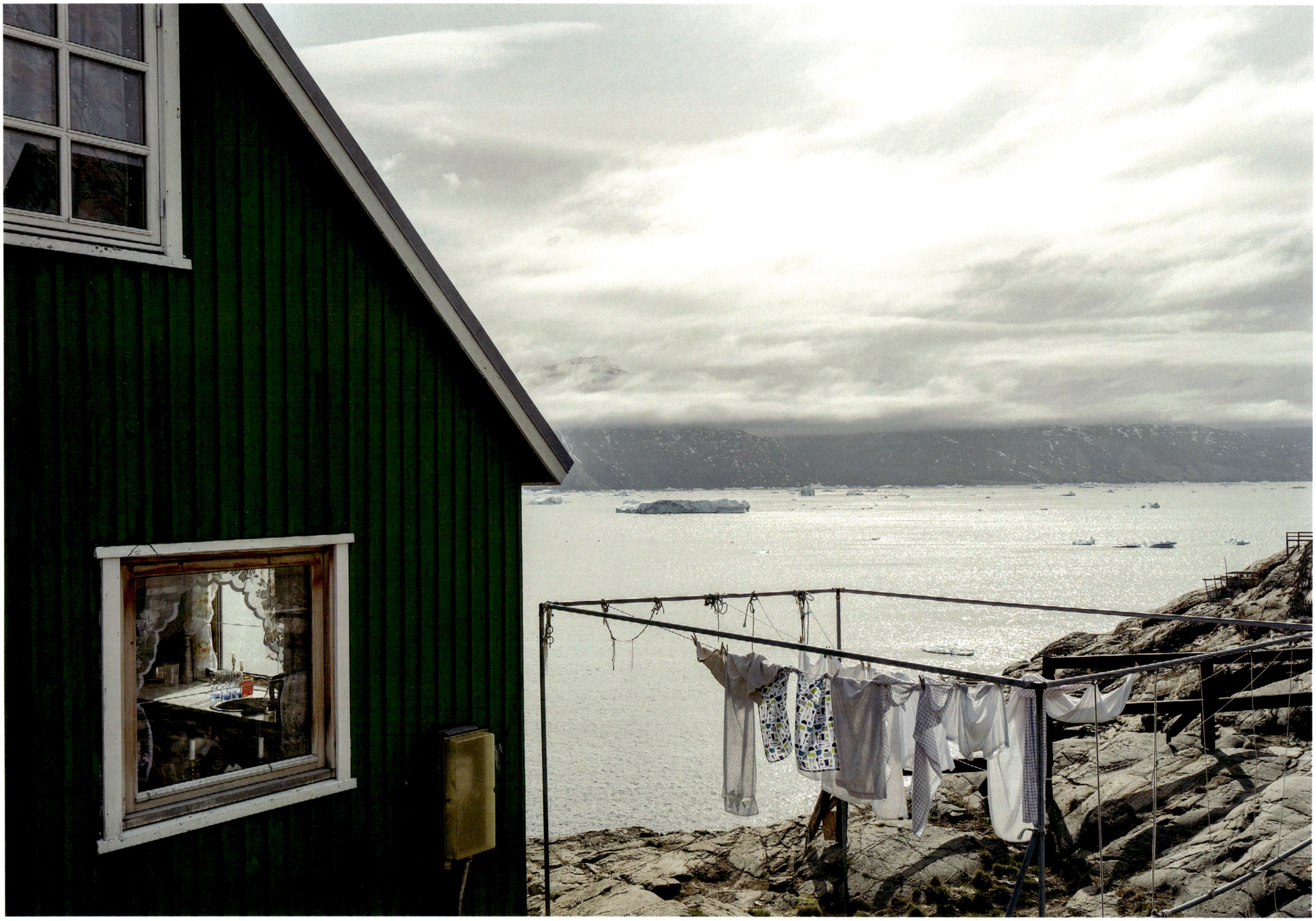

UUMMANNAQ 2016
Almost every home in Greenland has an amazing view of water and mountains.

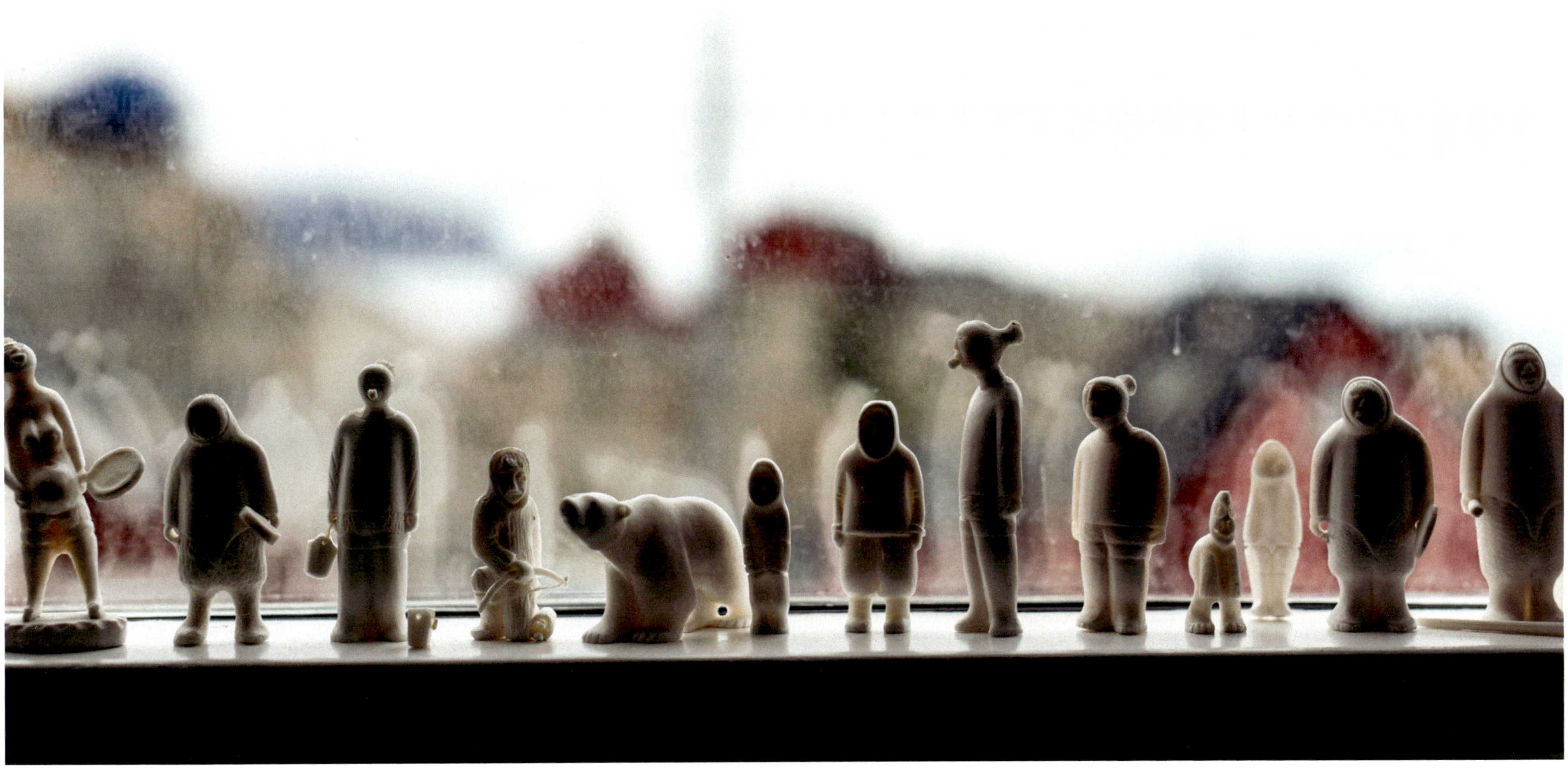

UUMMANNAQ 2016
Hand-carved walrus tusk figurines
from Ann Andreasen's collection.

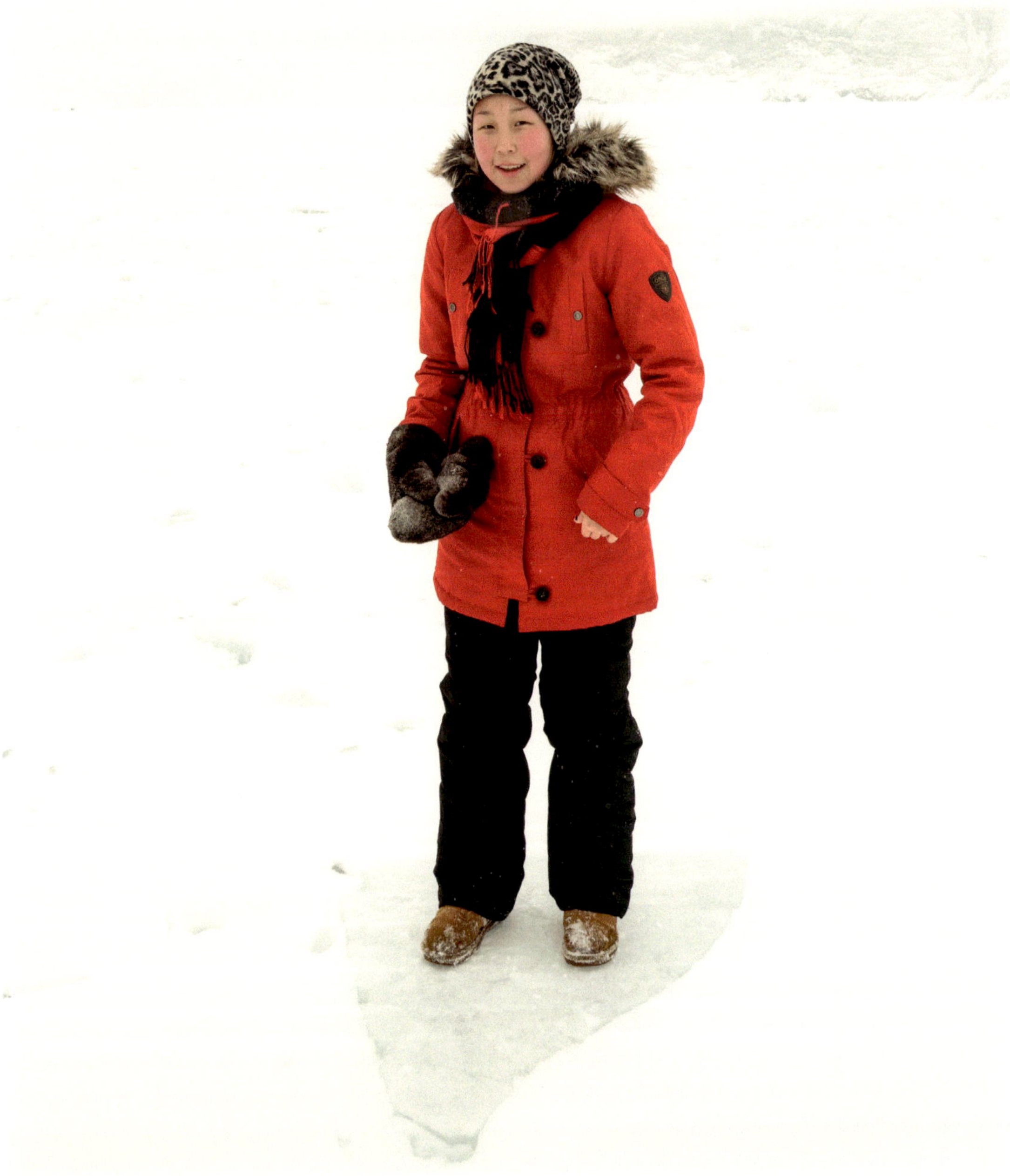

UUMMANNAQ 2017
Mona Mathiassen on the frozen Uummannaq Fjord.

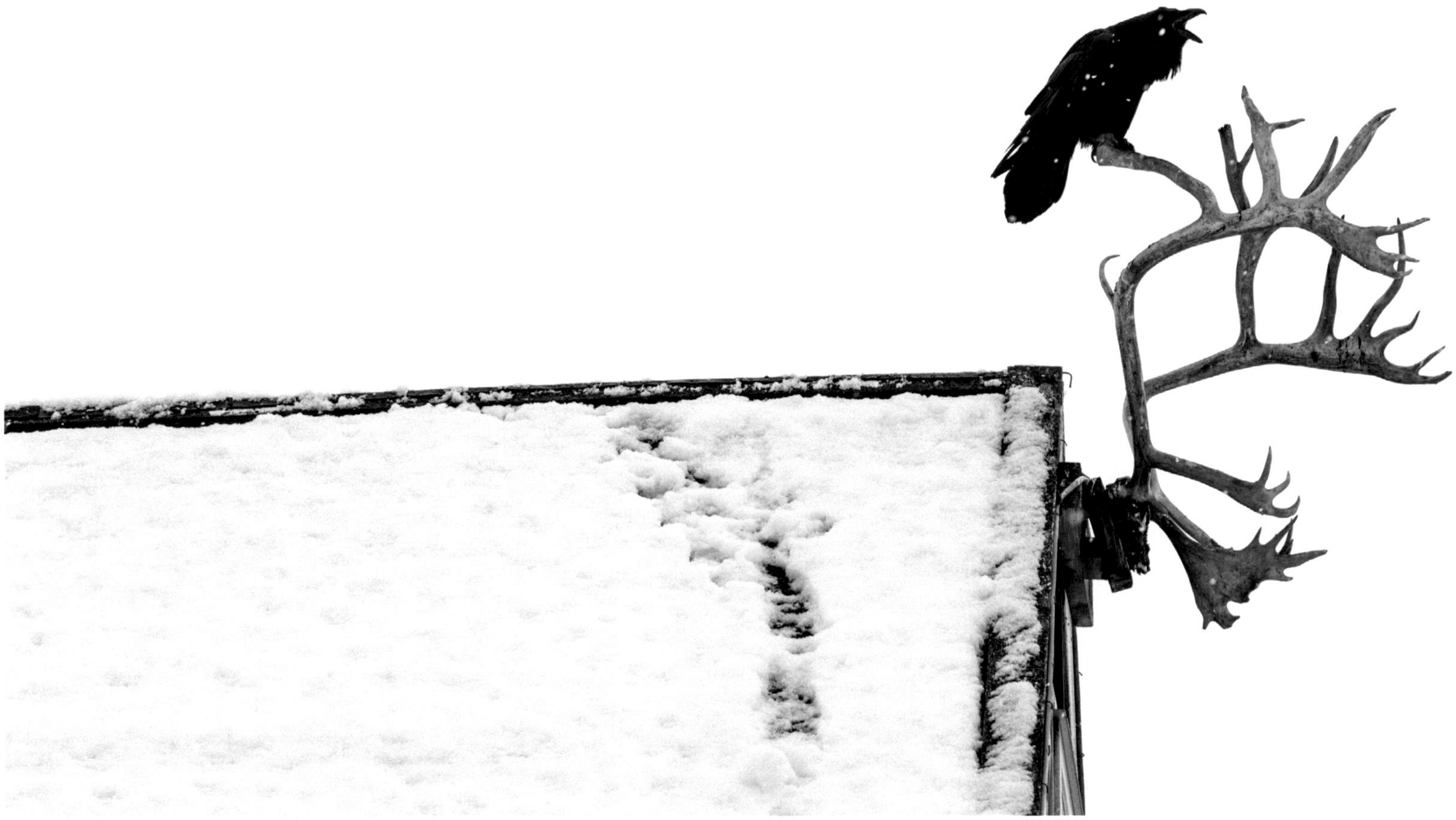

ILLORSUIT 2016
The raven caws to let all know
he has arrived.

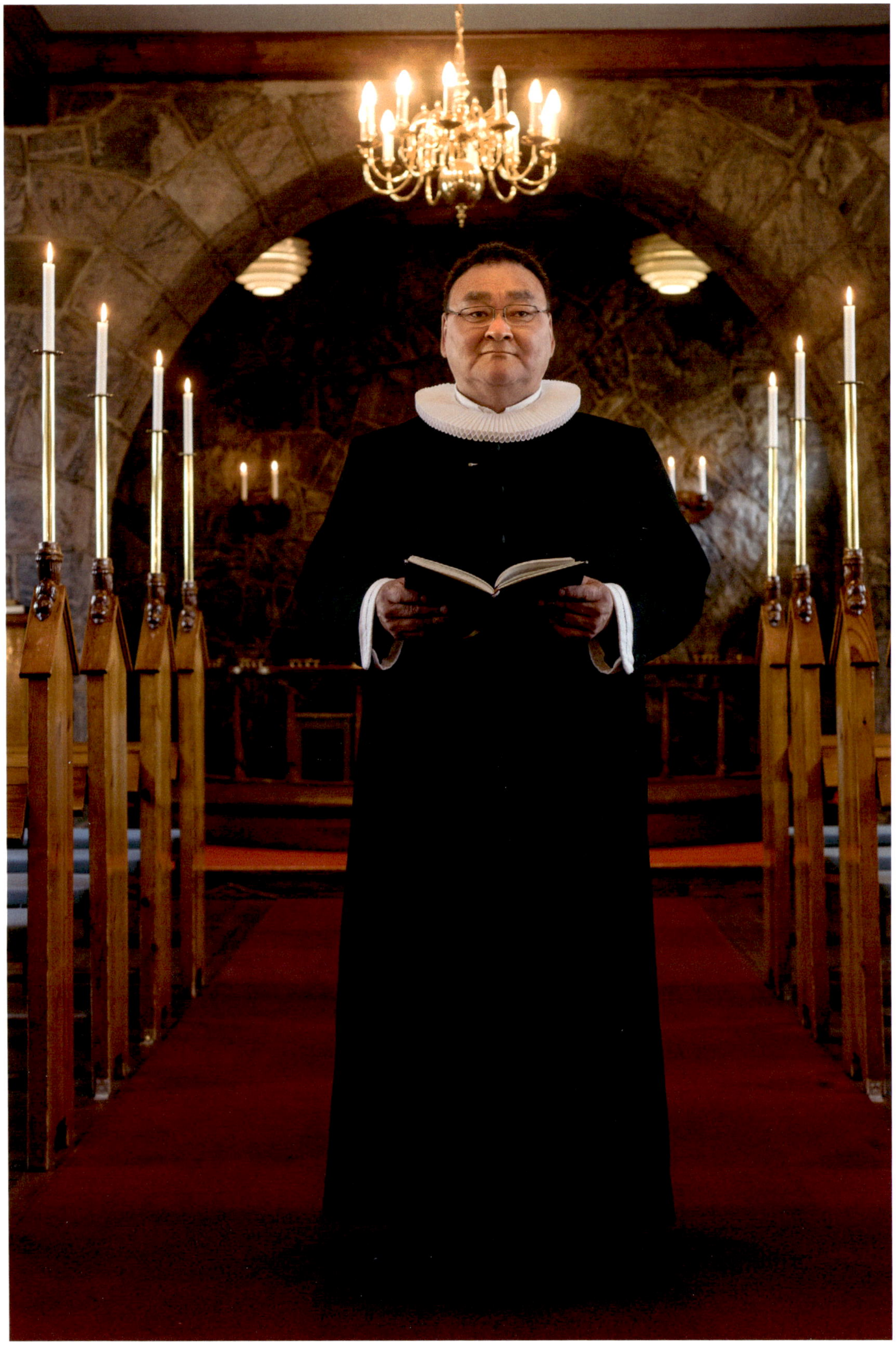

UUMMANNAQ 2016
Loogki Fleischer, parish priest. The religion of Greenland is derived from Lutheran and Moravian Brethren beliefs.

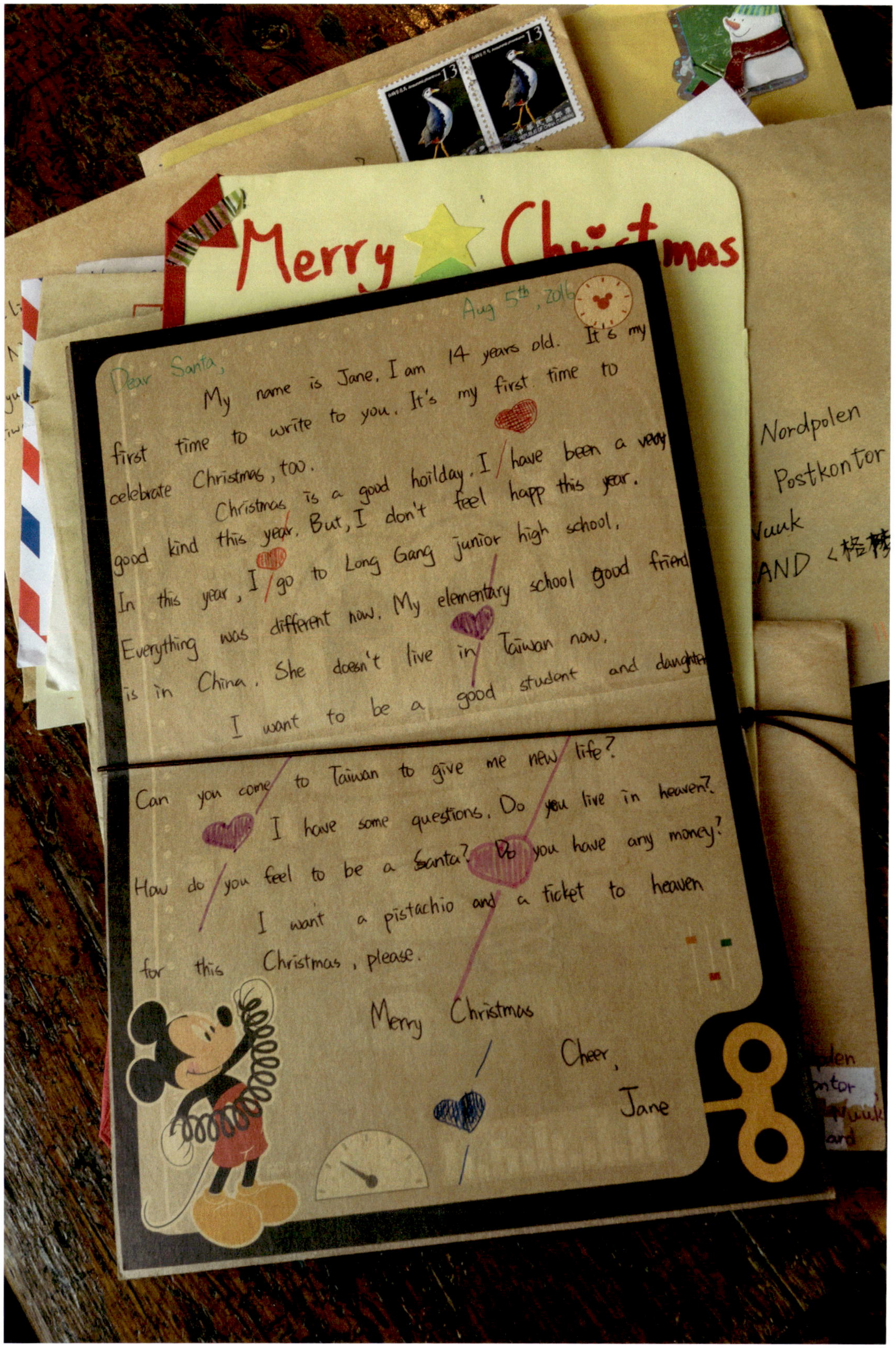


UUMMANNAQ 2017

Thousands of letters to Santa Claus from all over the world arrive each winter in Uummannaq, Santa's "summer home."

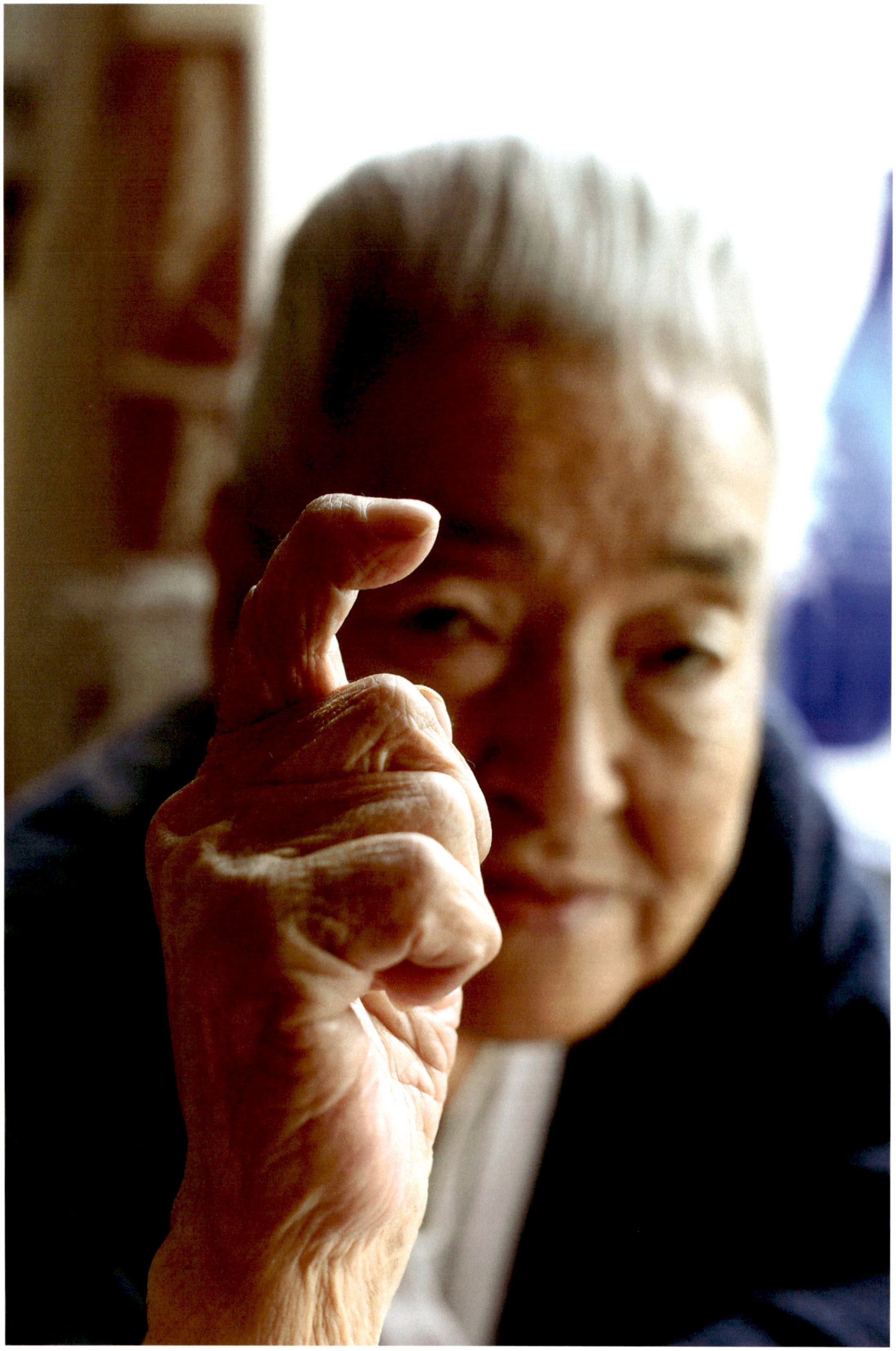

UUMMANNAQ 2016
Dorthe Simionsin shows her deformed finger, caused by a life of sewing seal skins.

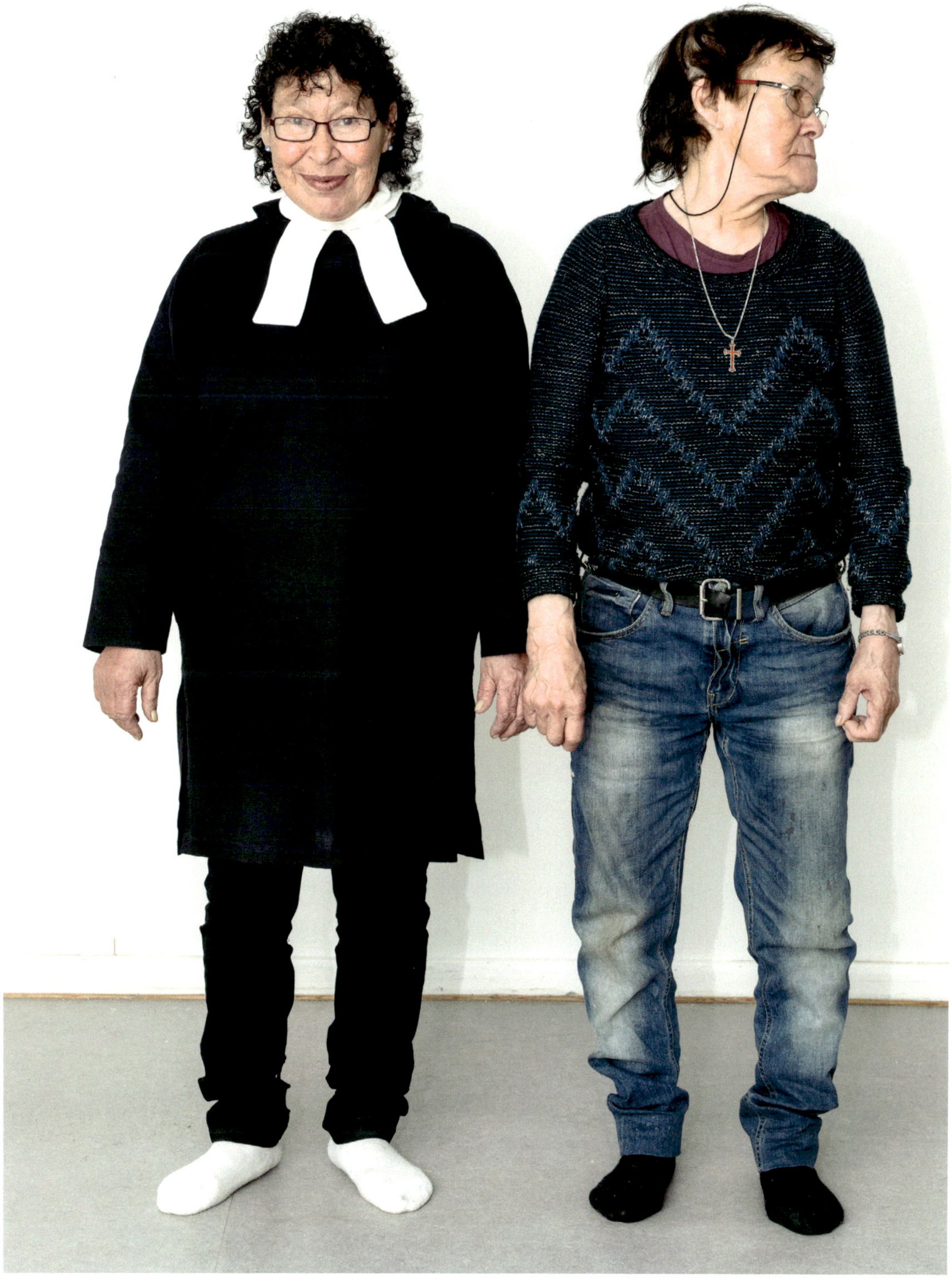

ILLORSUIT 2016
Portrait of Andrea Tobiassen
and Julianne Qvist.

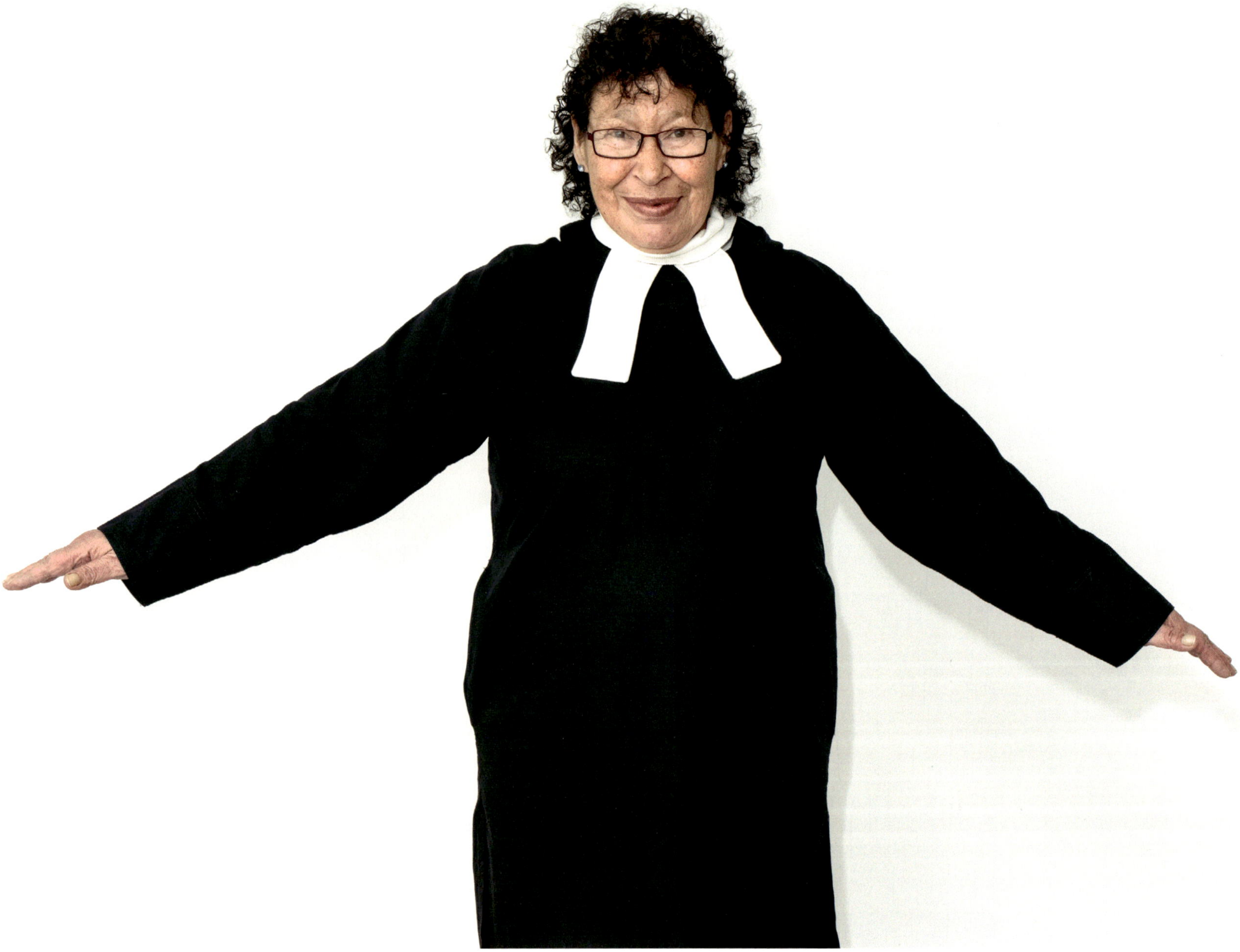

ILLORSUIT 2016
Portrait of Andrea Tobiassen.

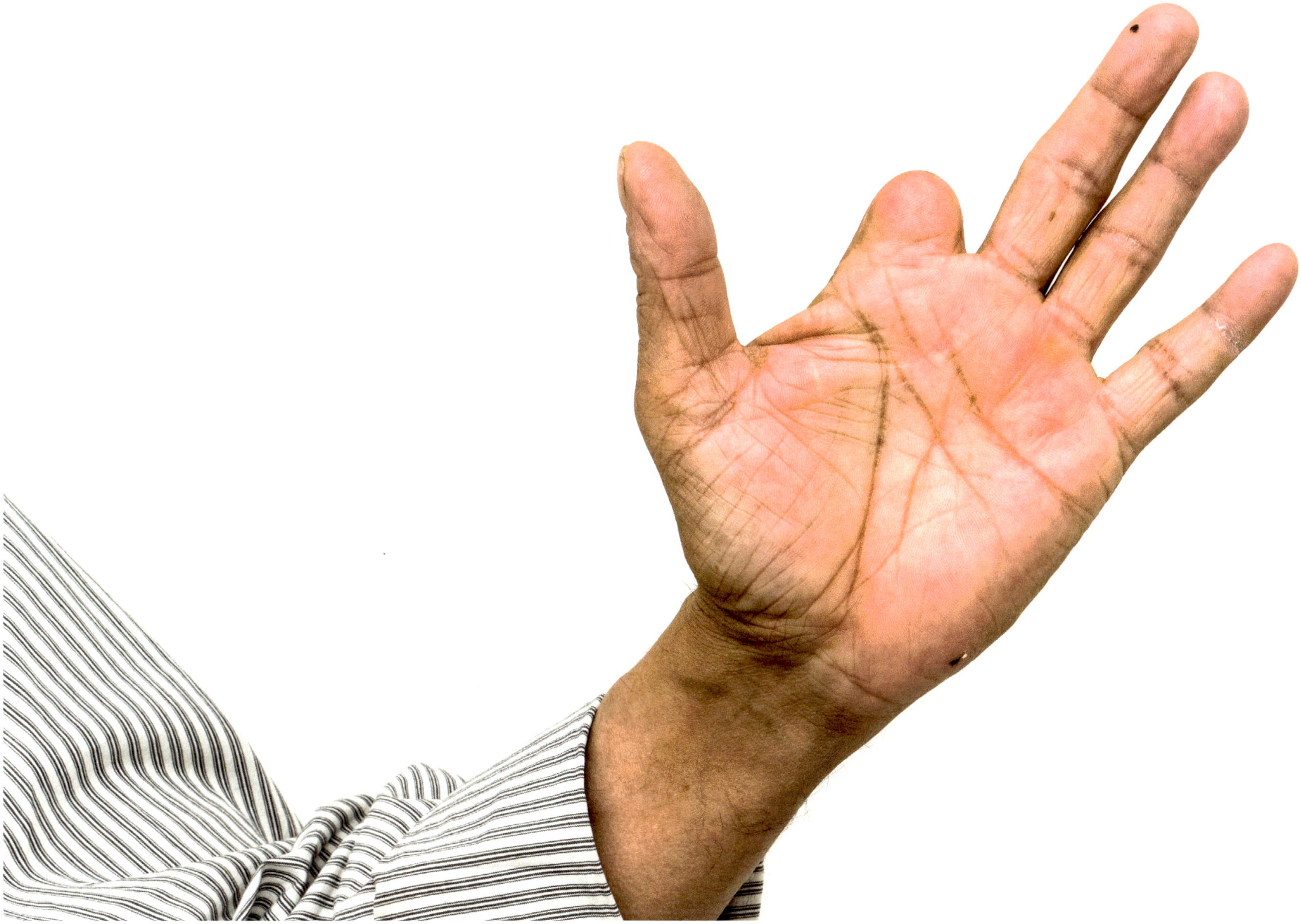

ILLORSUIT 2016
Portrait of Benjamin Jonaghansen's finger.

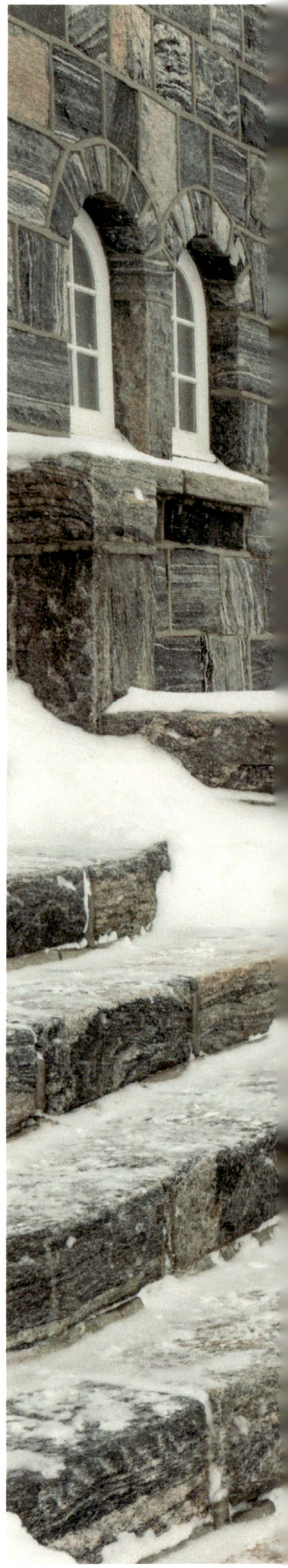

UUMMANNAQ 2017
Funeral procession.

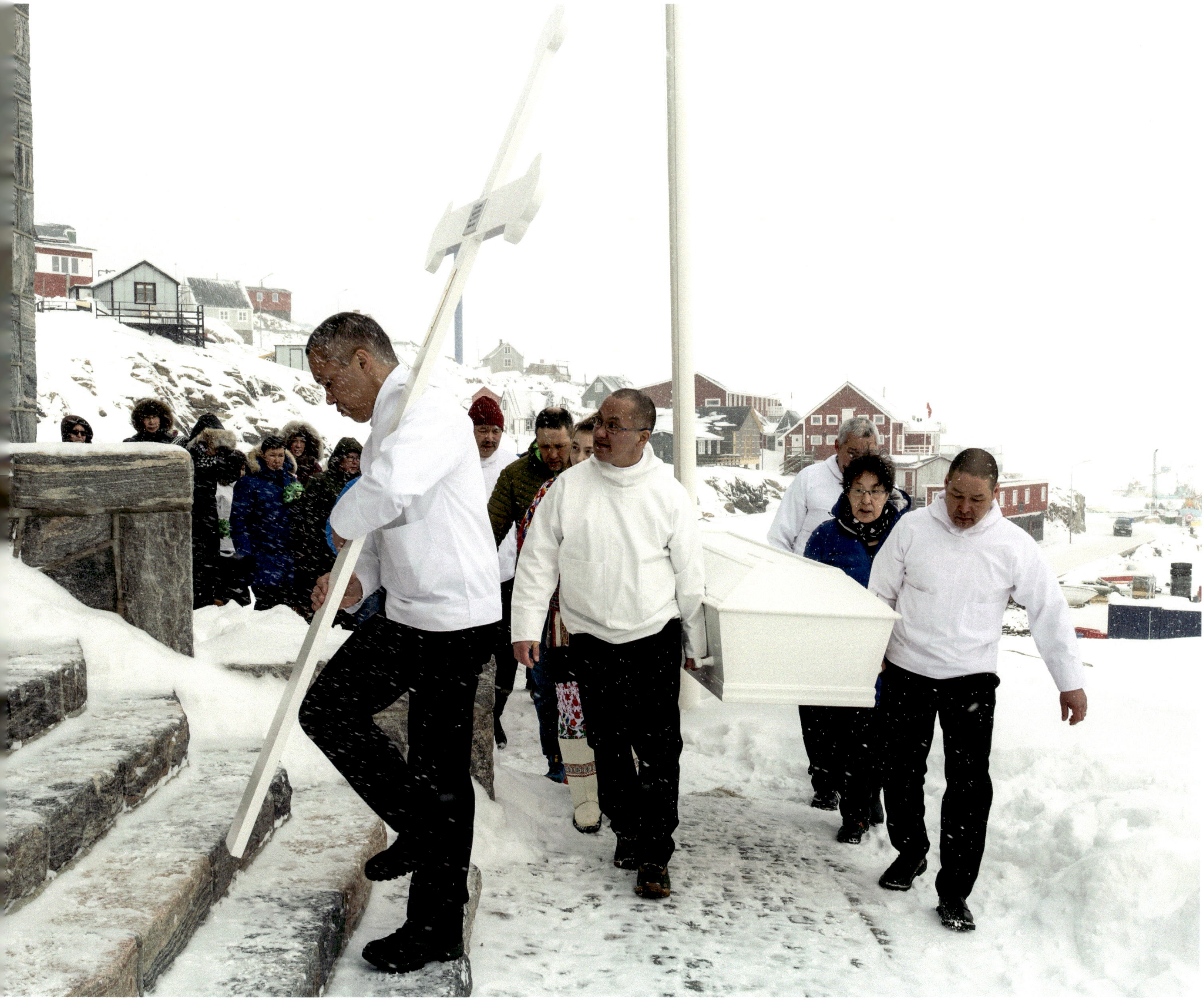

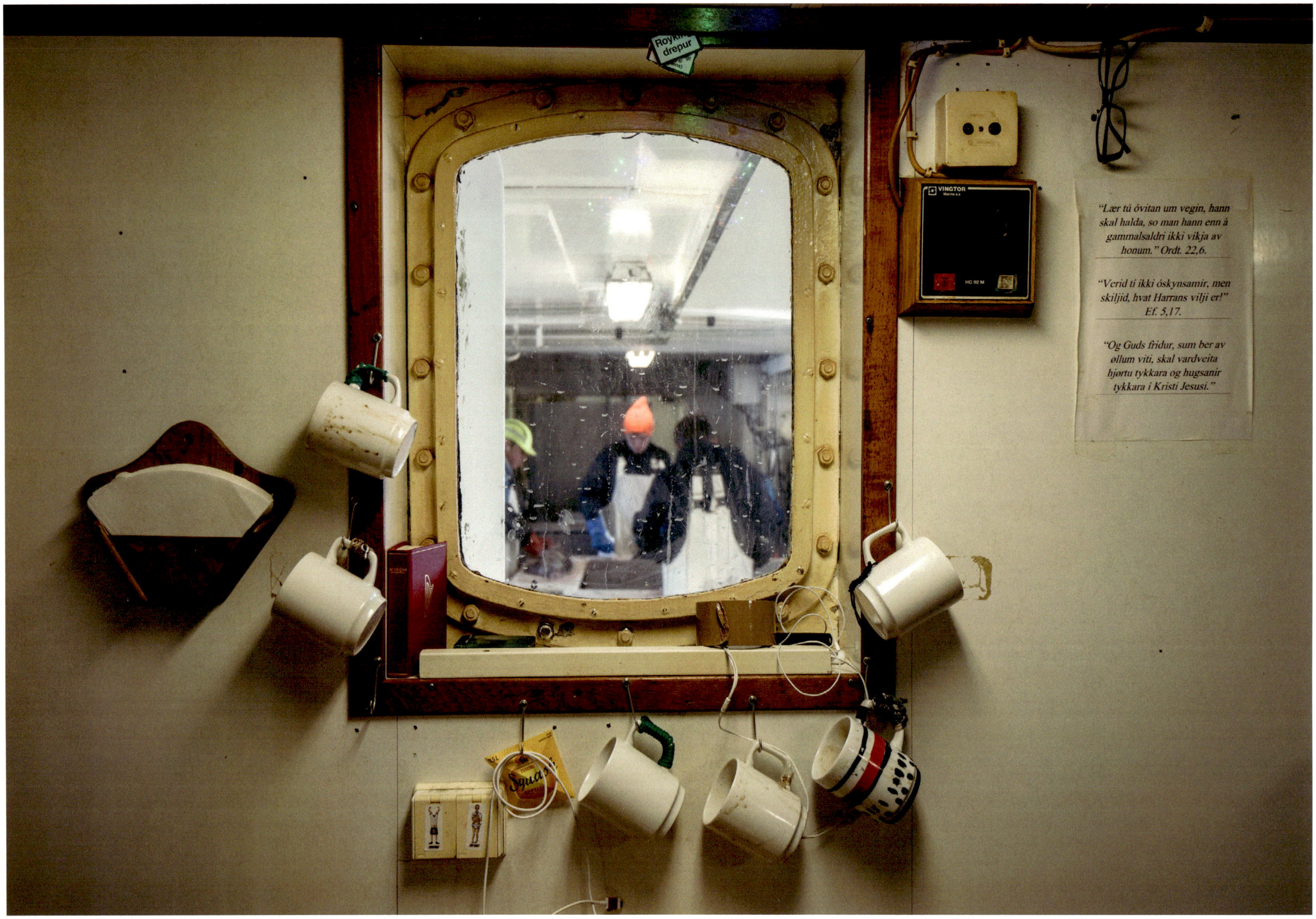


UUMMANNAQ DISTRICT 2017

View of the fish processing area and breakroom of the Faroe Island-based ship, *Agot*.

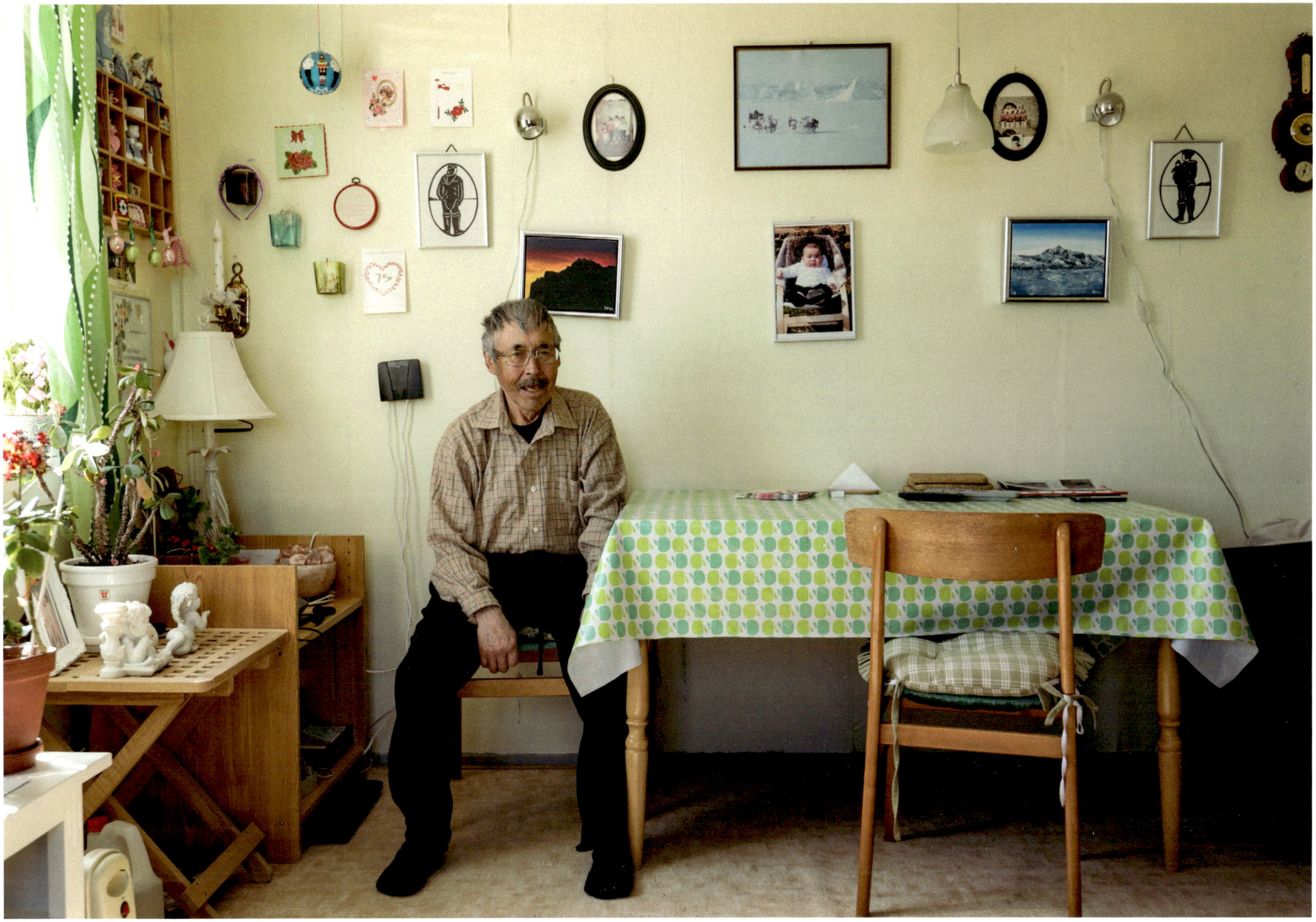

ILLORSUIT 2016

Long-time hunter Aron Nielson in his living room. "Silence was all there was in the past hunting from kayaks and dog sledge," Aron explains.

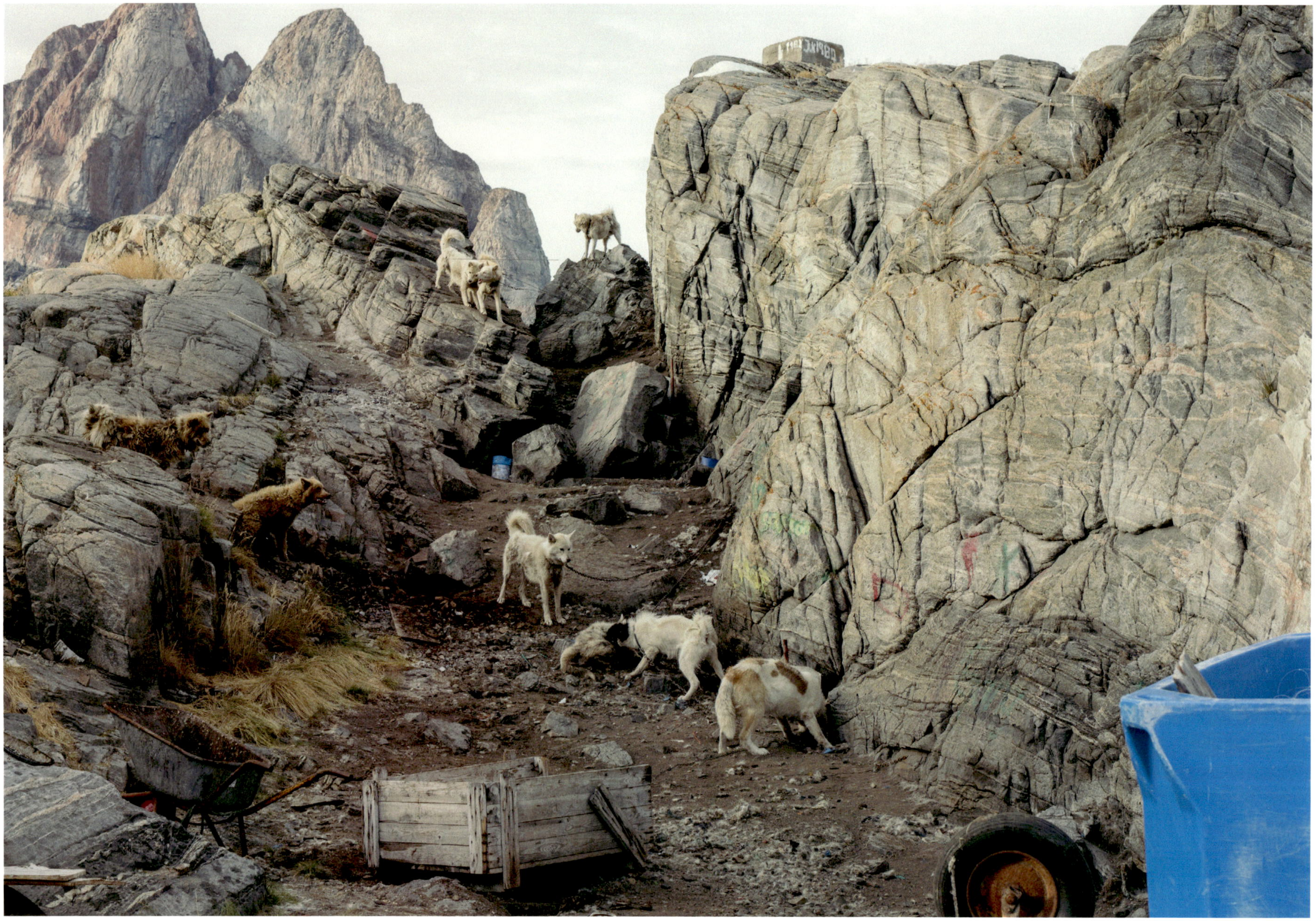

UUMMANNAQ 2017
Chained dogs.

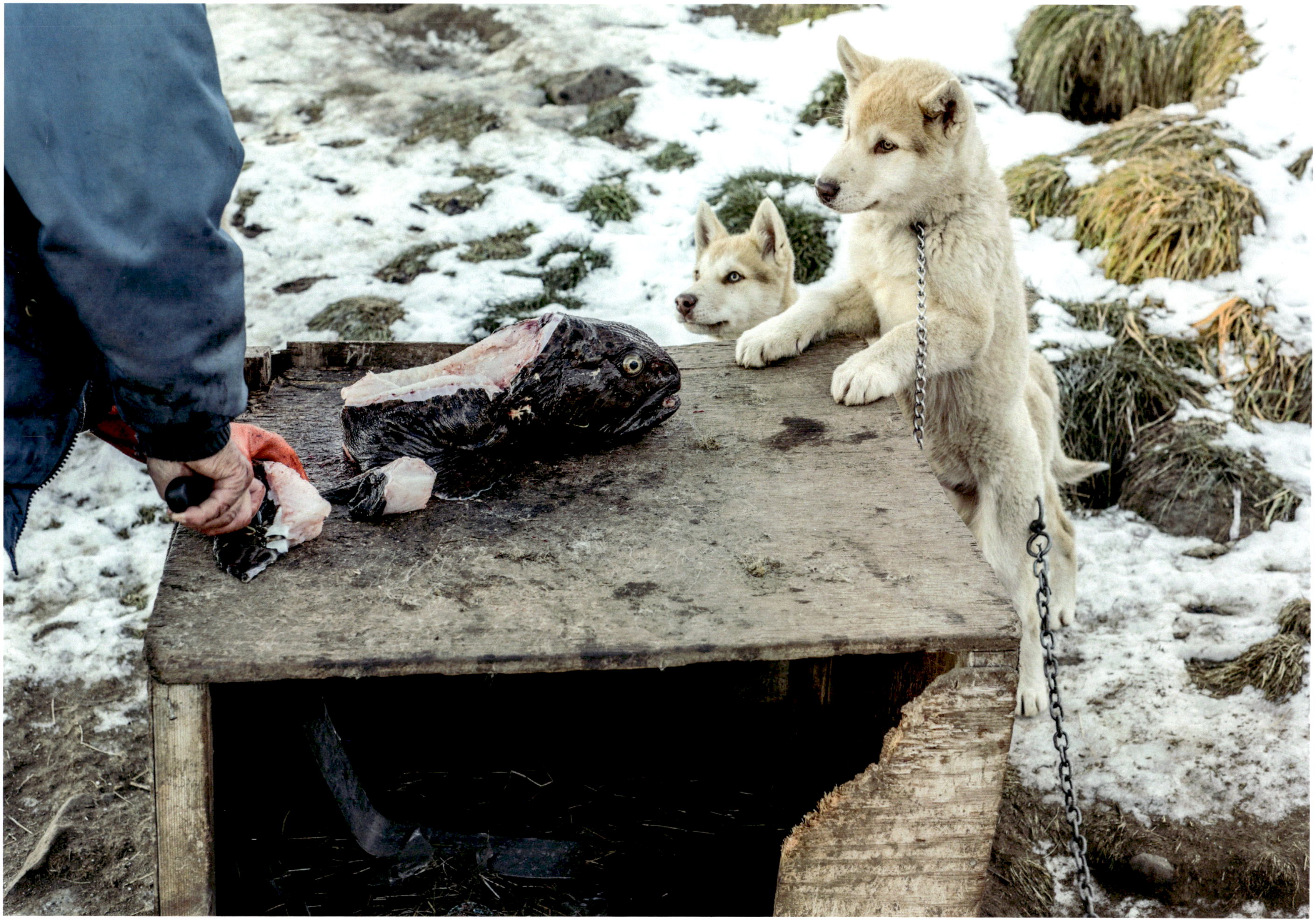

UUMMANNAQ 2016
Excited puppies waiting for dinner.

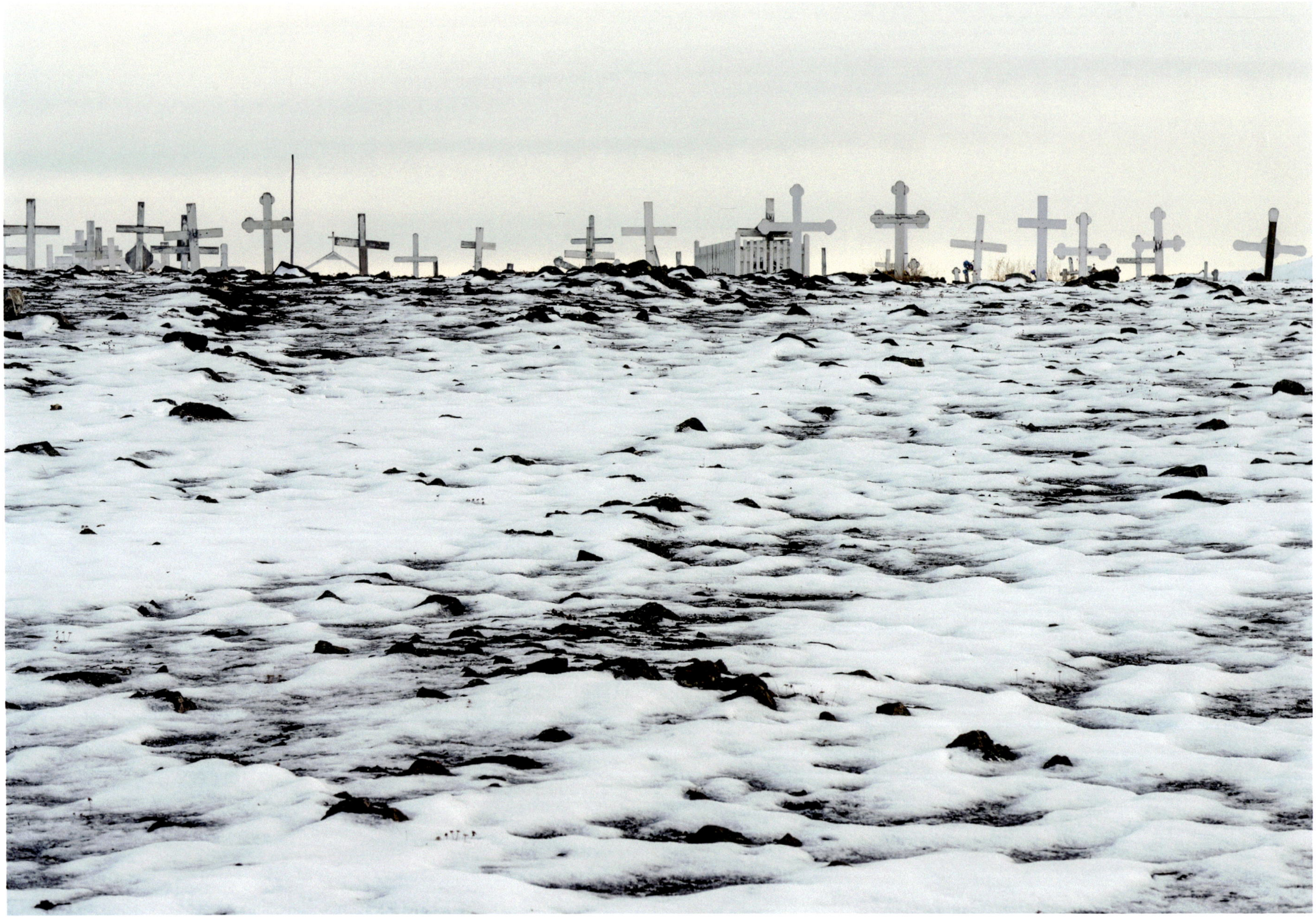

ILLORSUIT 2016

Illorsuit cemetery on the southern headlands.

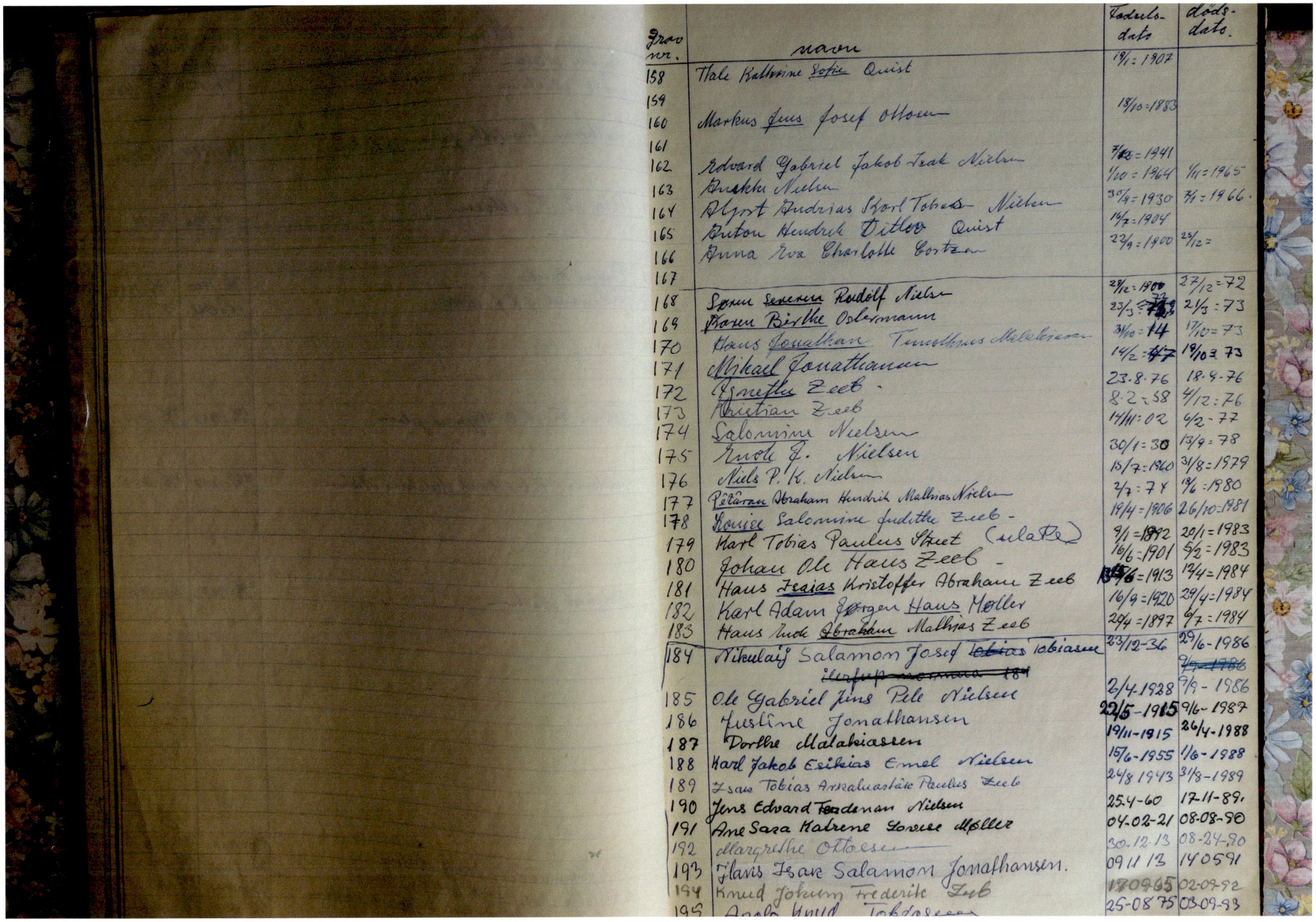

ILLORSUIT 2016

Birth and death journal ledger containing names of many people whom Kent wrote about in his book, *Salamina*.

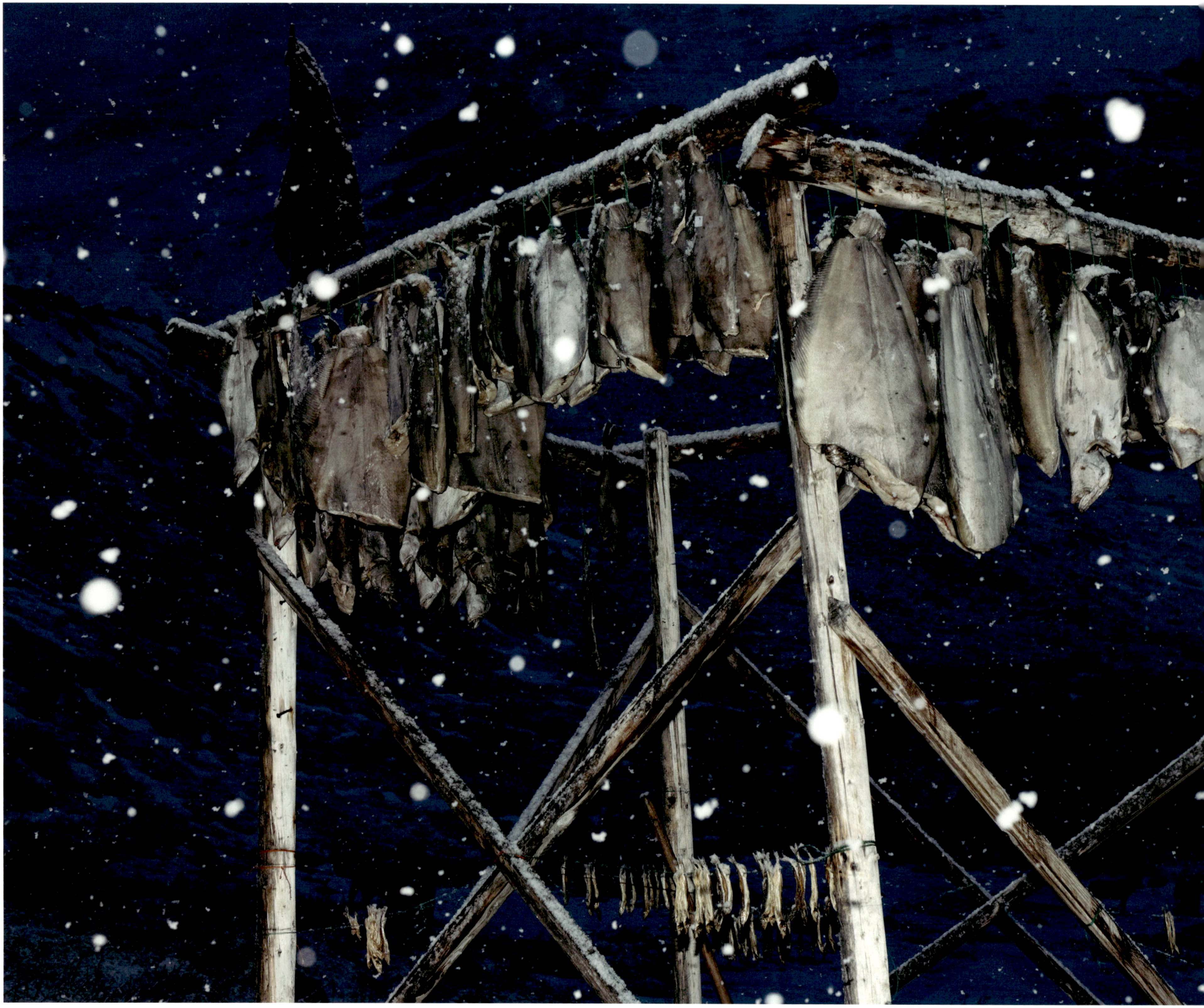

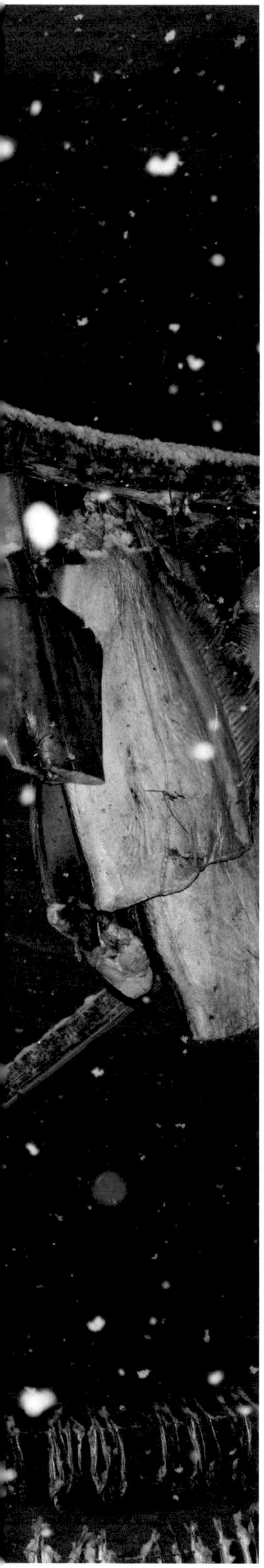

ILLORSUIT 2017
Fish hang on traditional drying racks on a cold winter's night. Halibut and capelin will dry over time in the Arctic air.

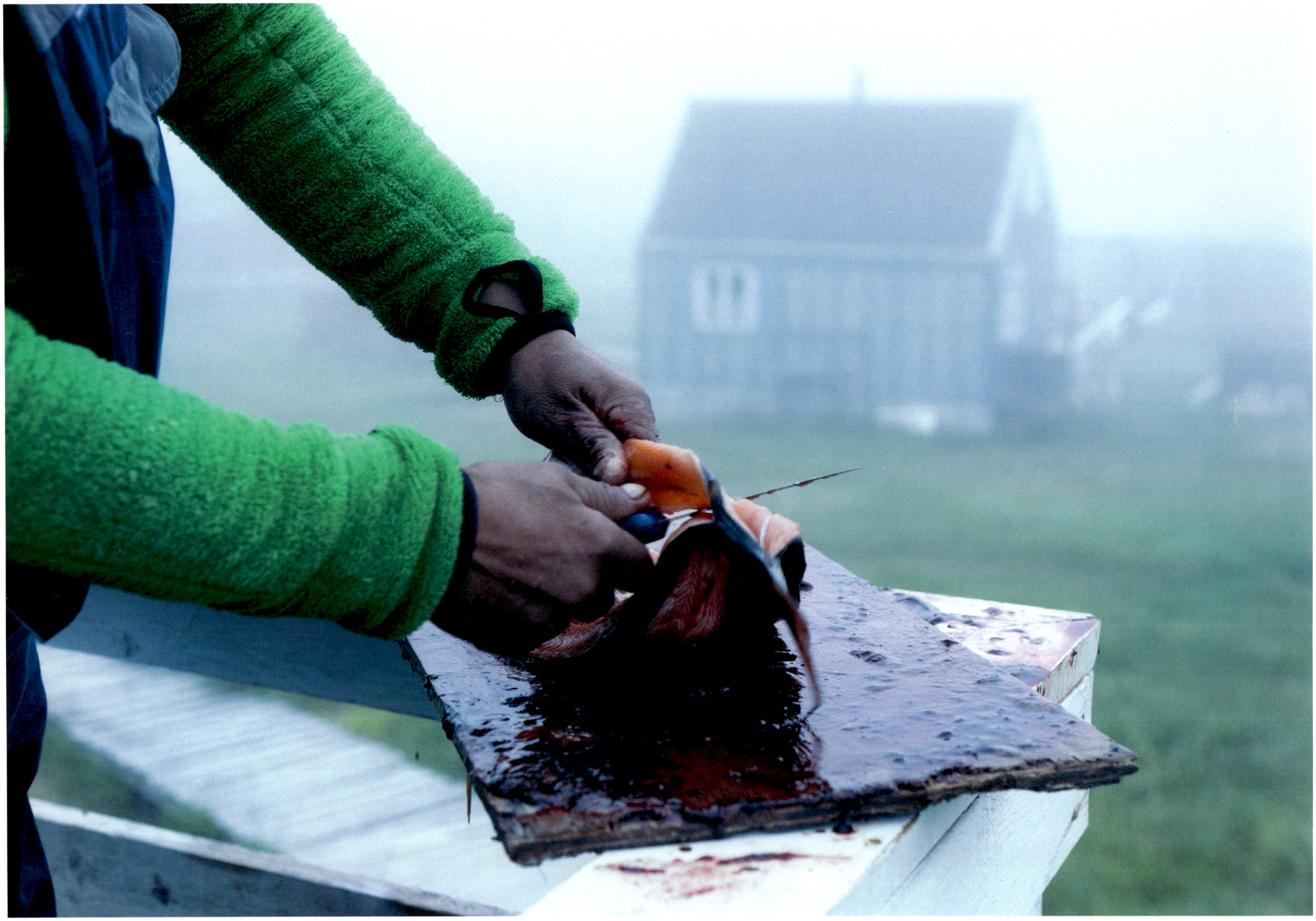

ILLORSUIT 2016
Filleting salmon in the midnight fog.

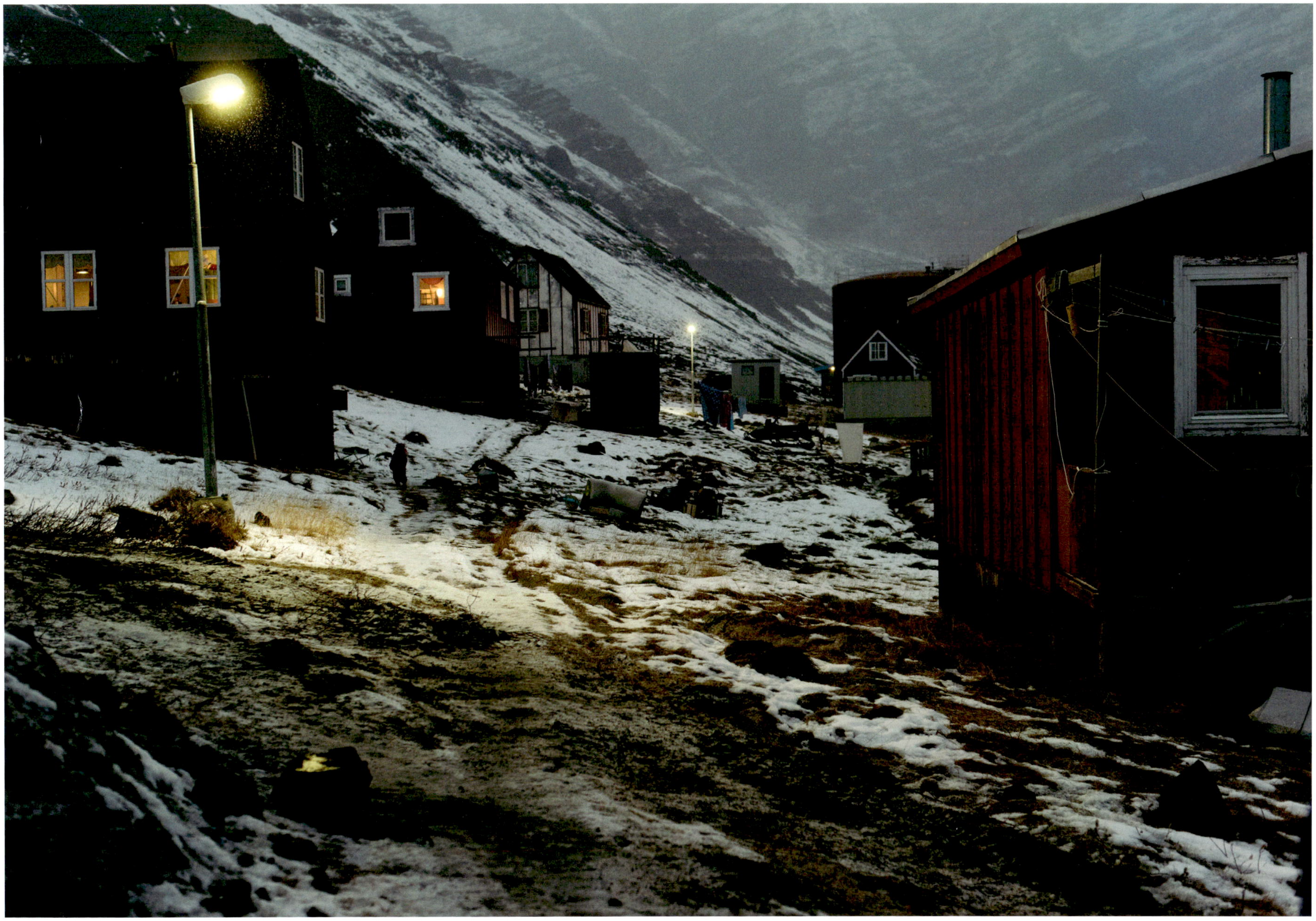

ILLORSUIT 2016
Cold winter night in the settlement.

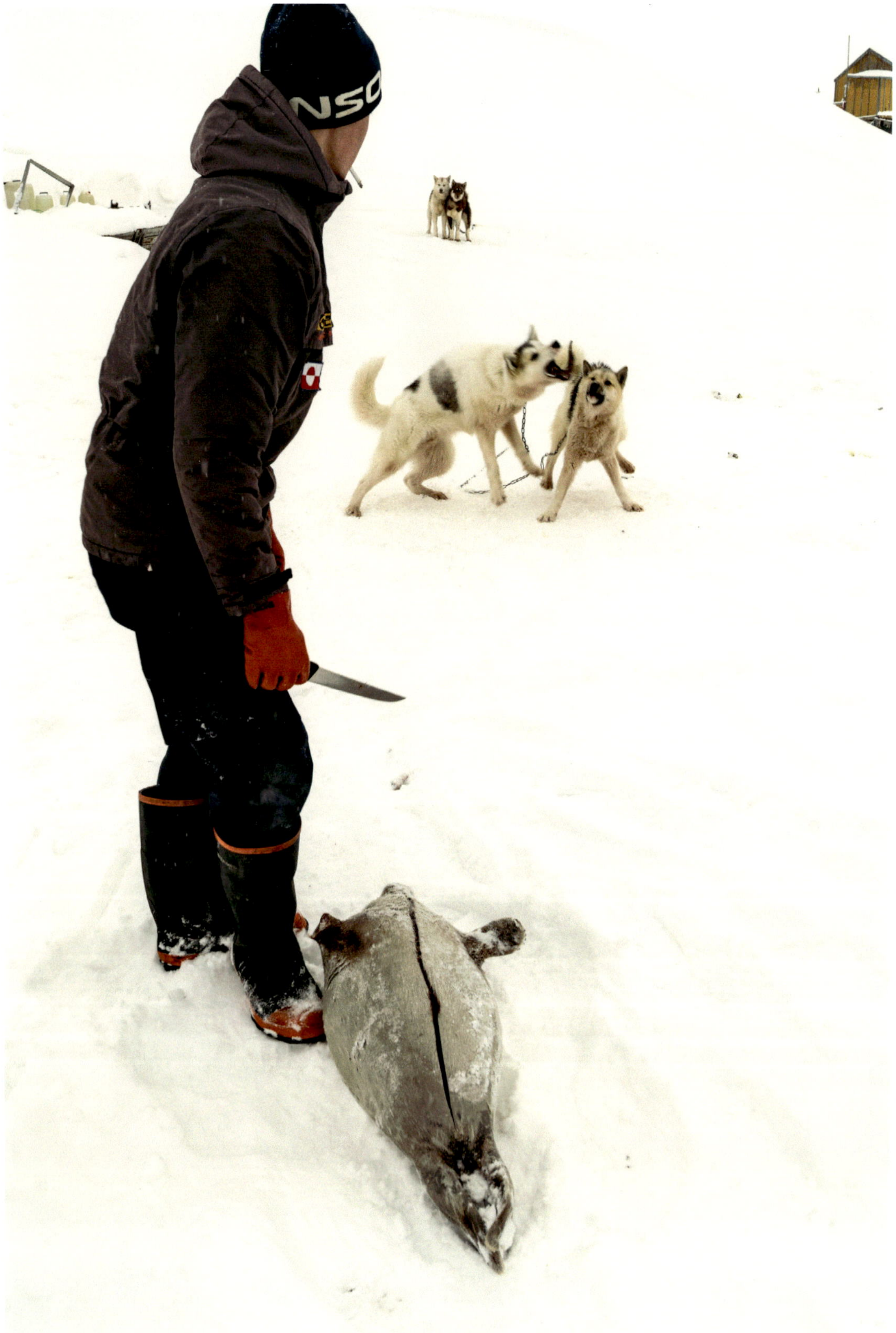


ILLORSUIT 2017
Hungry dogs wait for food.

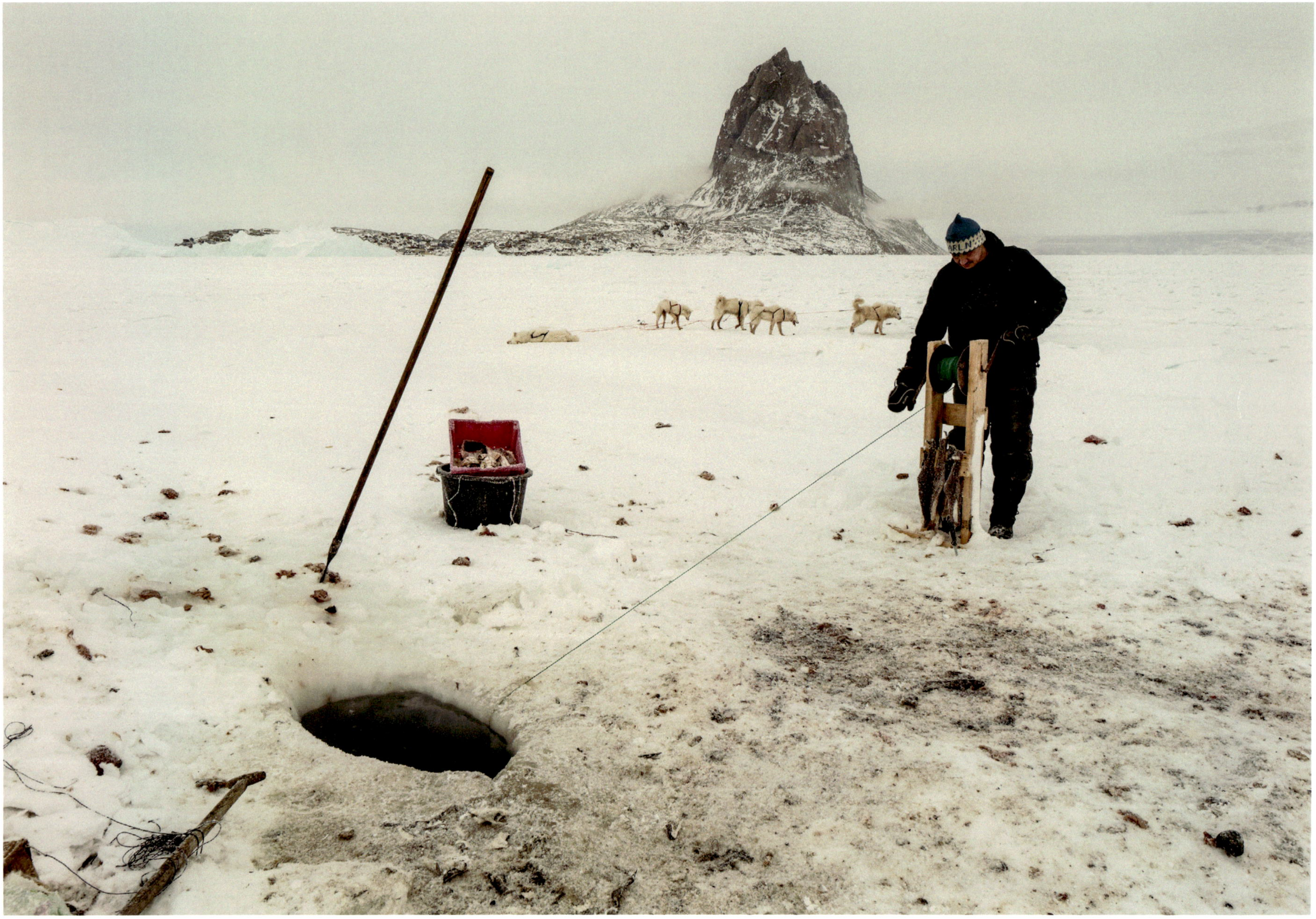

UUMMANNAQ 2017
After dog sledging to his fishing hole,
Karl Nielsen pulls in halibut on the long line.

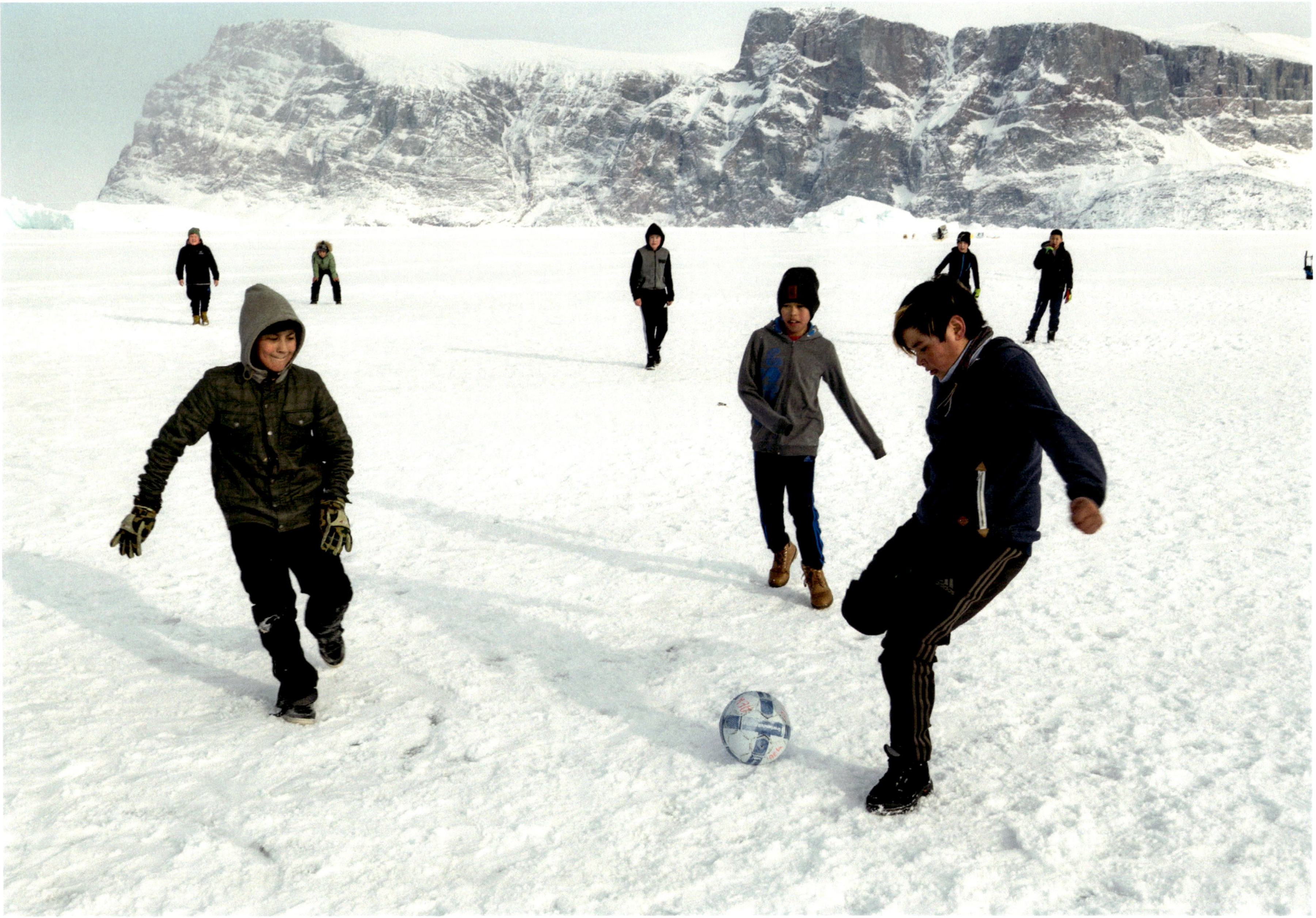

UUMMANNAQ 2017
Despite the slippery footing, soccer practice takes place on the frozen Uummannaq Fjord.

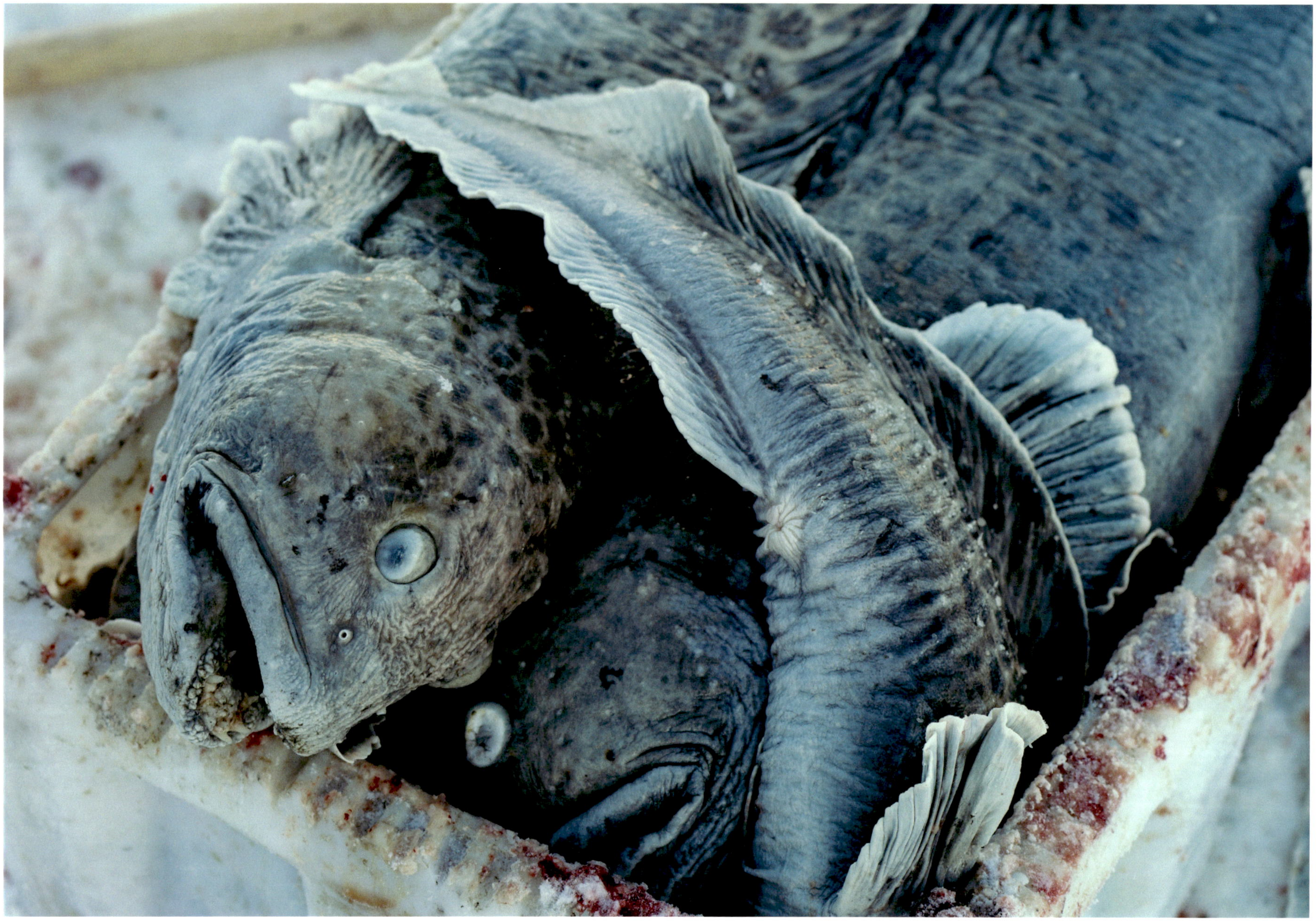

ILLORSUIT 2017
A box full of frozen wolffish
from the day's catch.

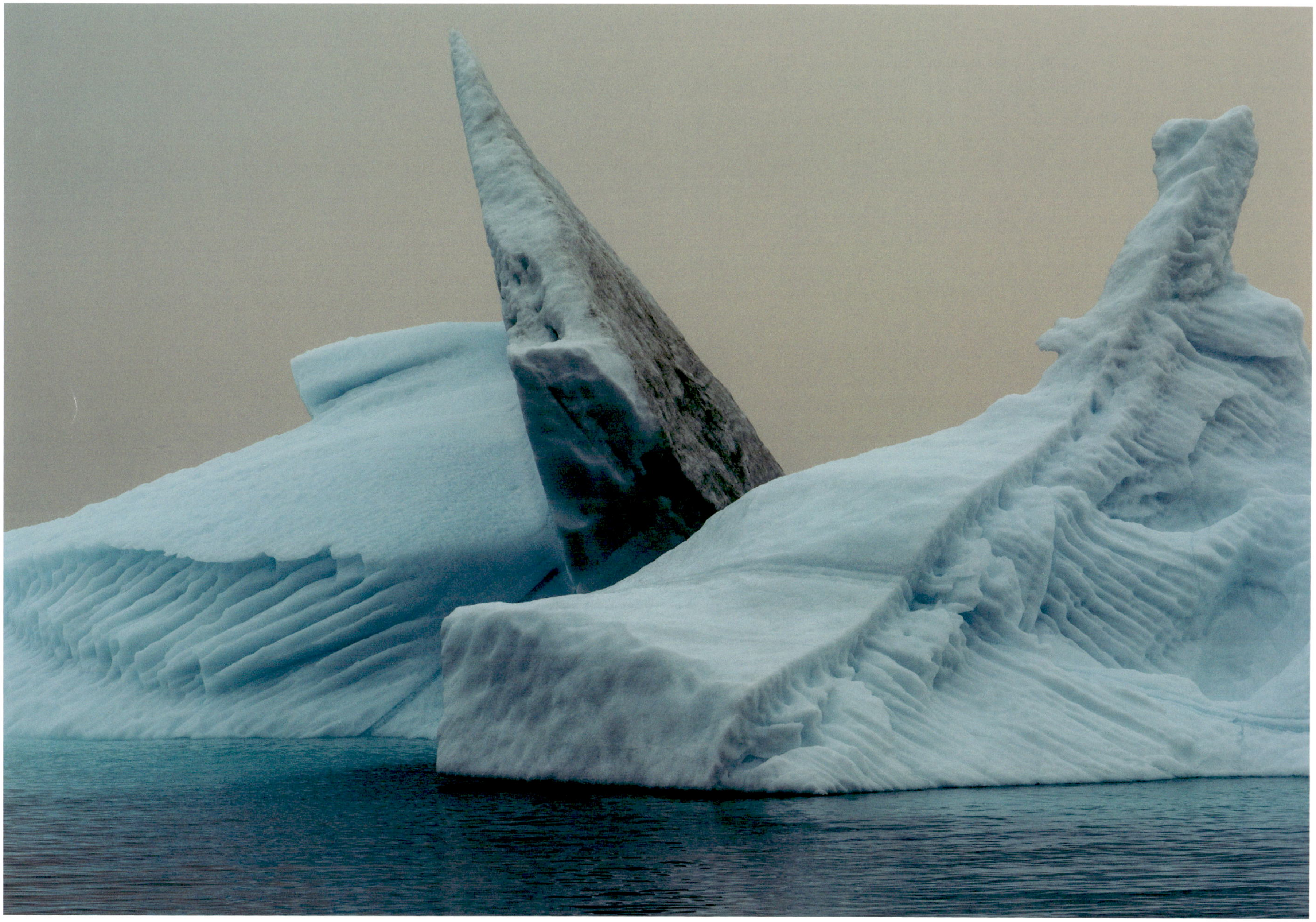

ILLORSUIT SOUND 2017
Iceberg before the snowstorm.

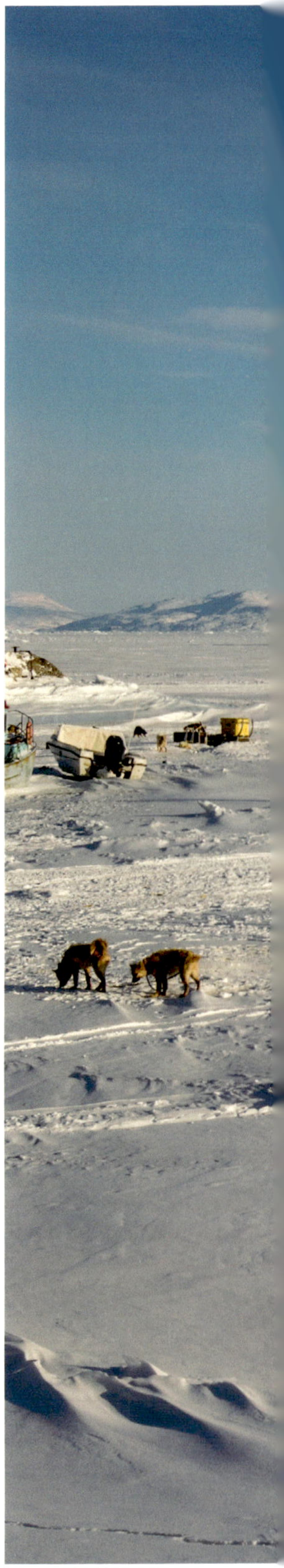

UUMMANNAQ 2017
Uummannaq Harbor, where the winter freeze is shared by boats and dogs.

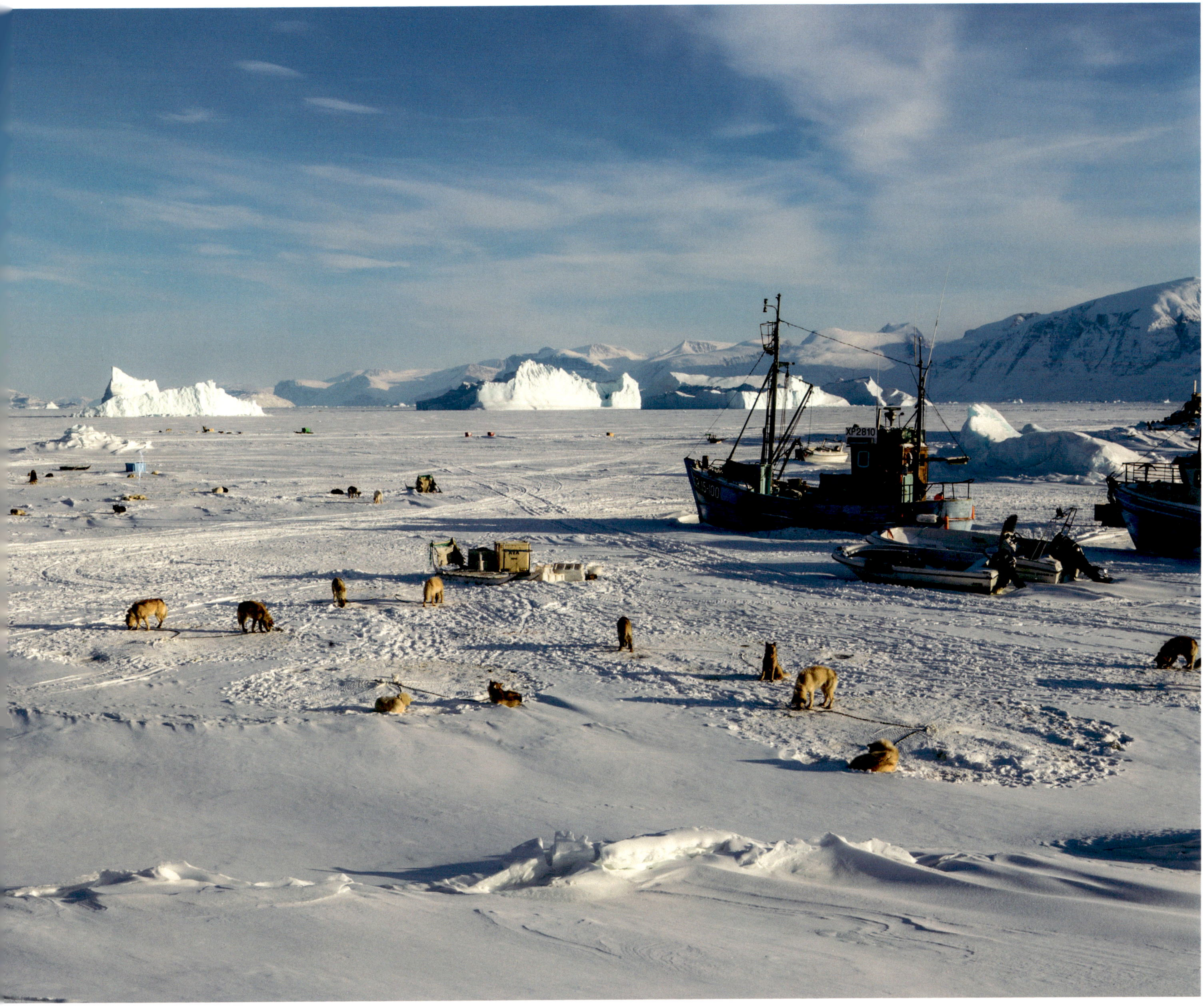
XP2810

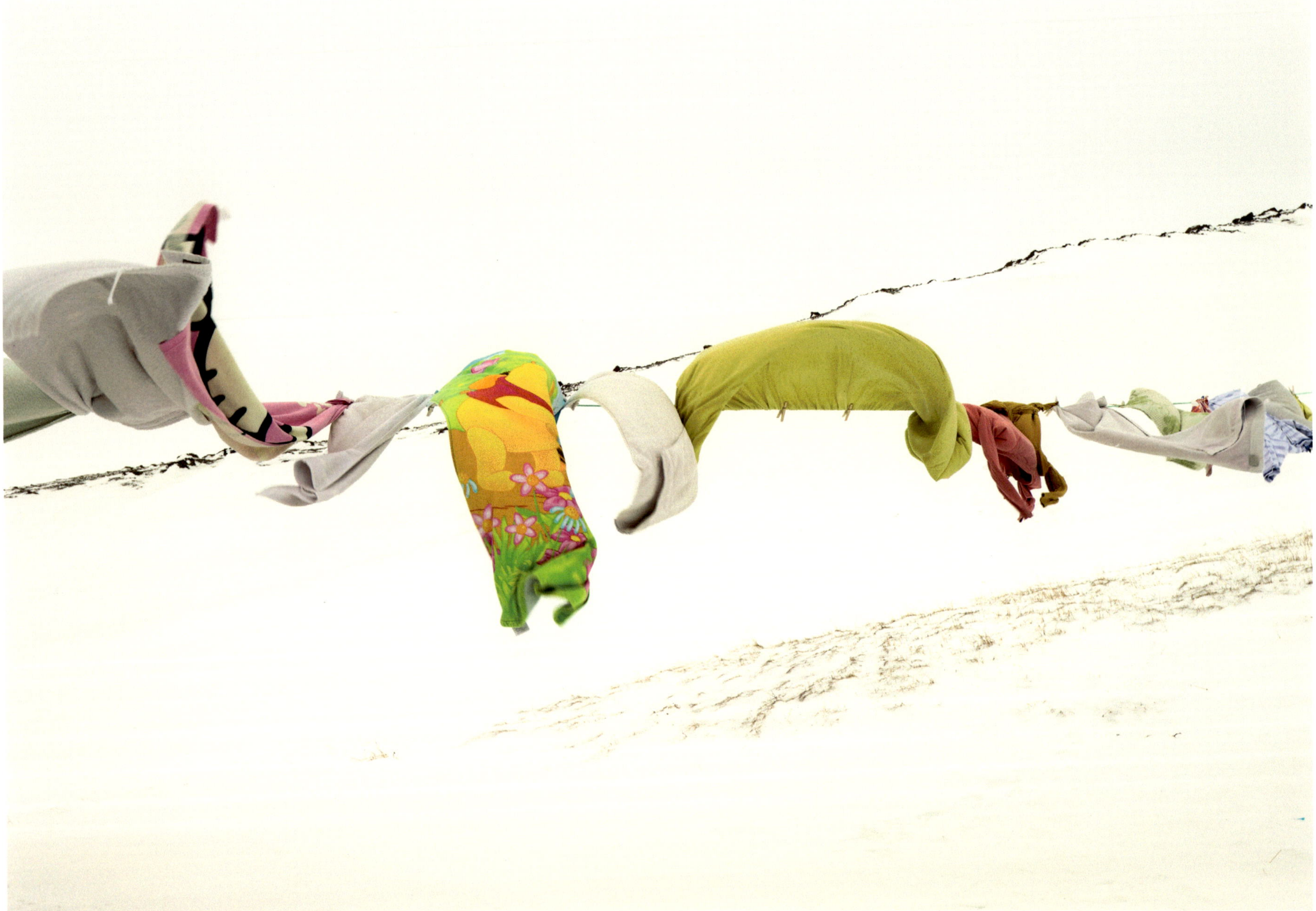

ILLORSUIT 2017

Clothes on a line dance in the winter wind.

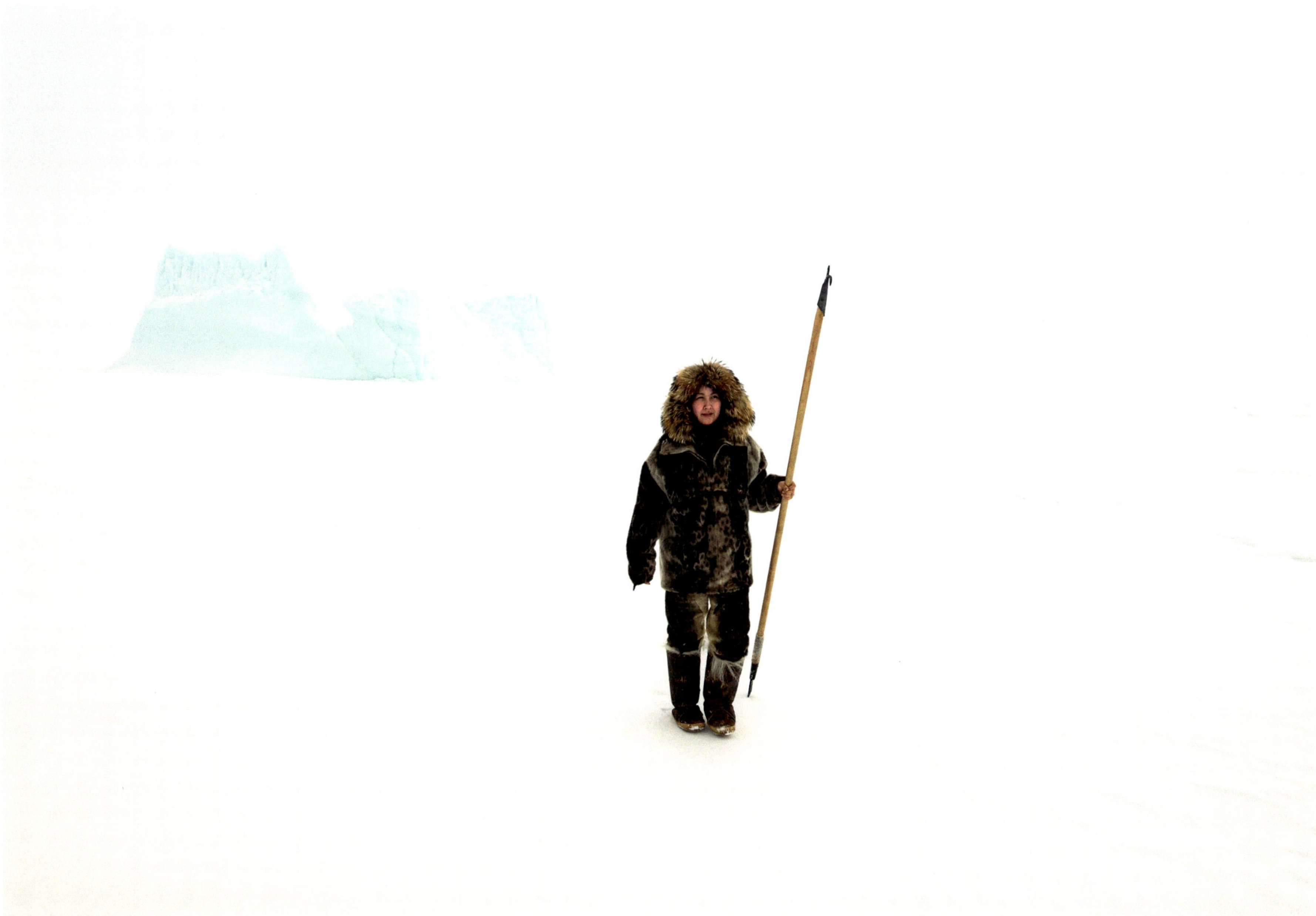

ILLORSUIT 2017

Sara Lyberth in her favorite seal skins.

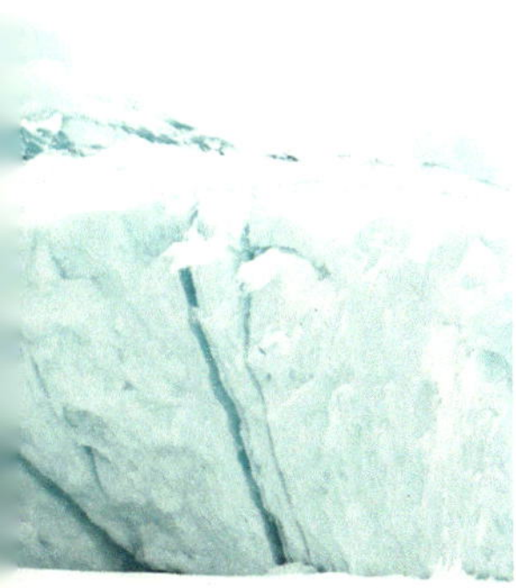

UUMMANNAQ 2017
Exploring icebergs on the frozen fjord.

UUMMANNAQ 2017
Looking through an iceberg bridge at another iceberg in the distance.

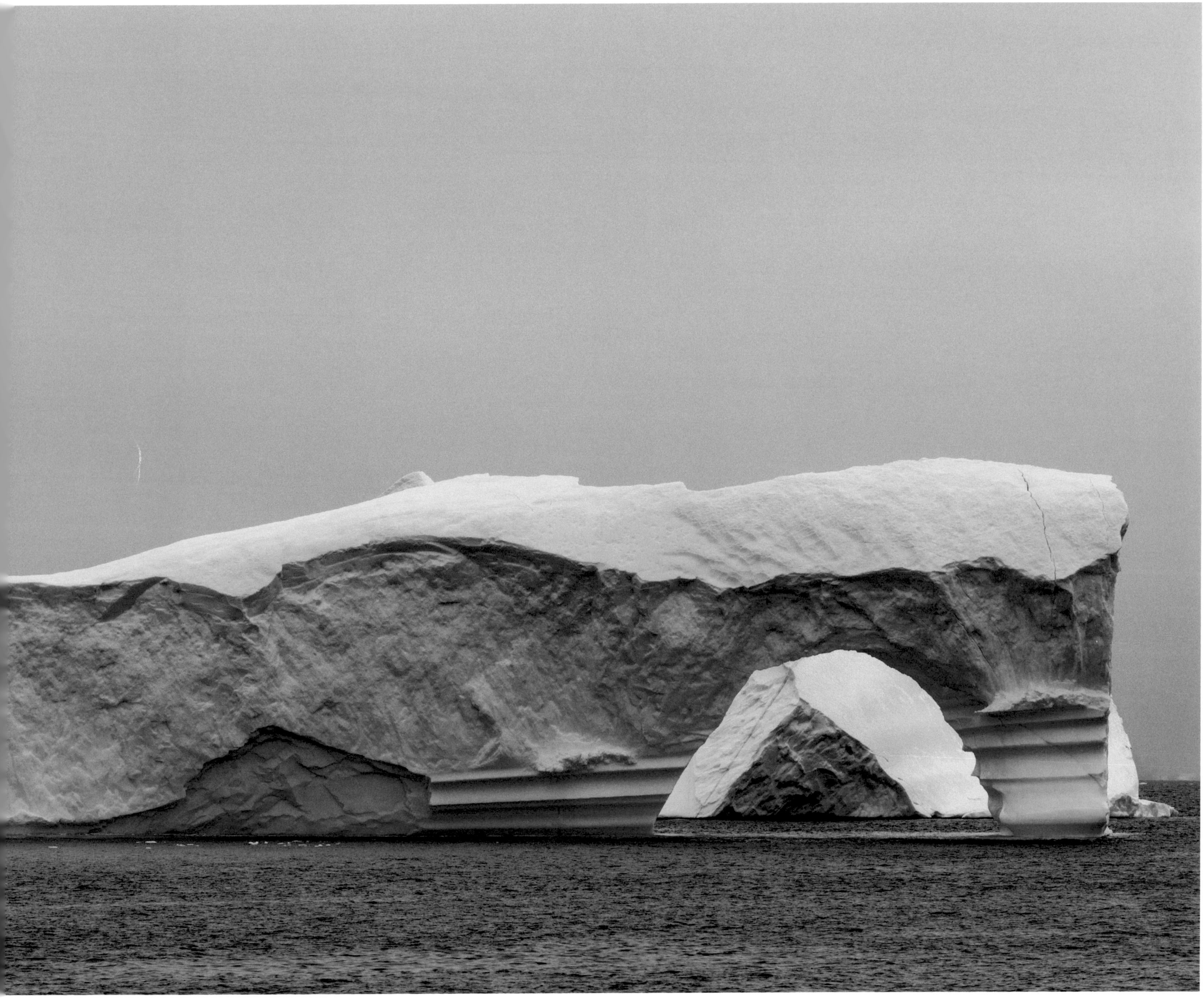

ILLORSUIT 2017
The night sky with Northern Lights over the path leading to the settlement's helipad.

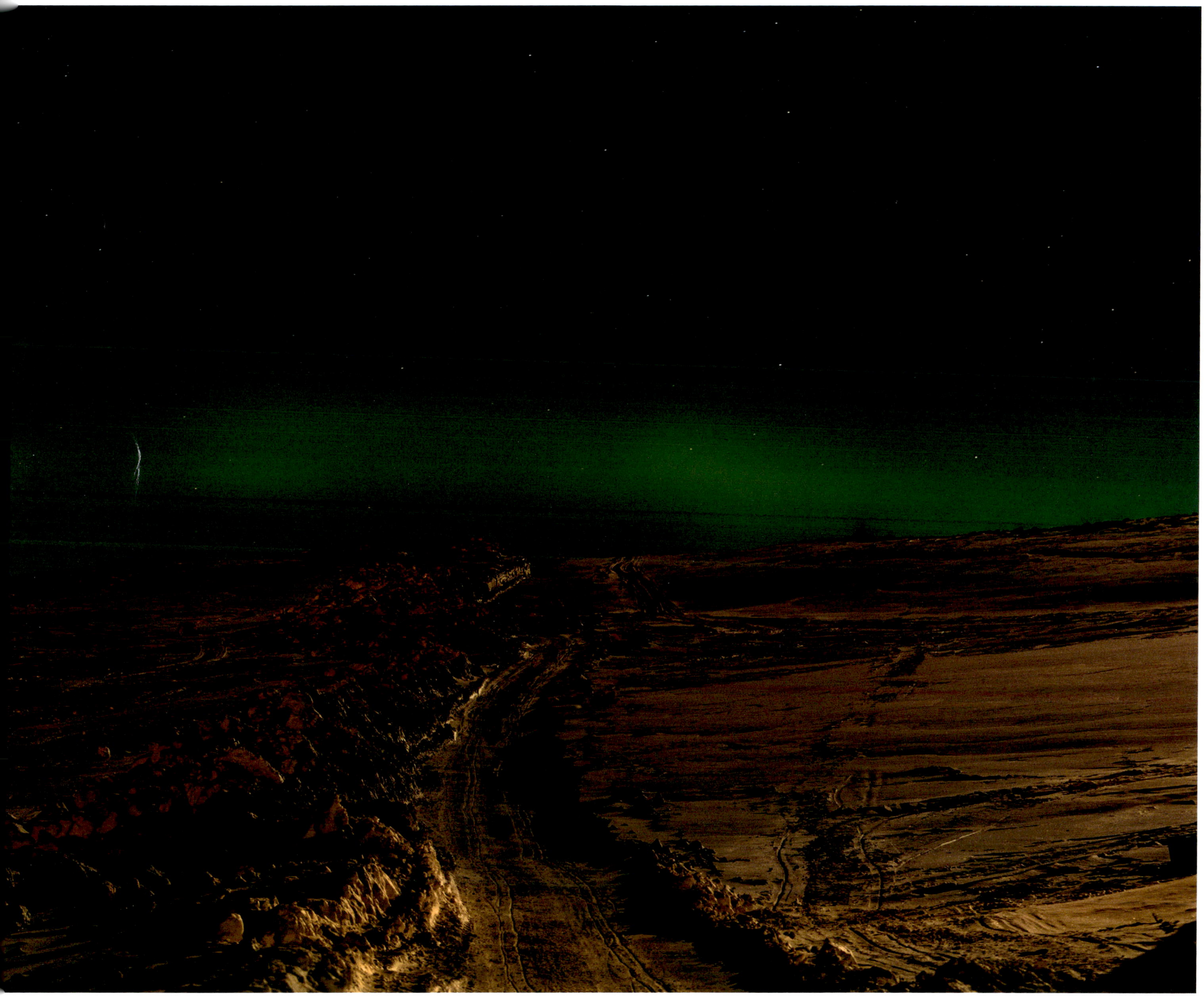

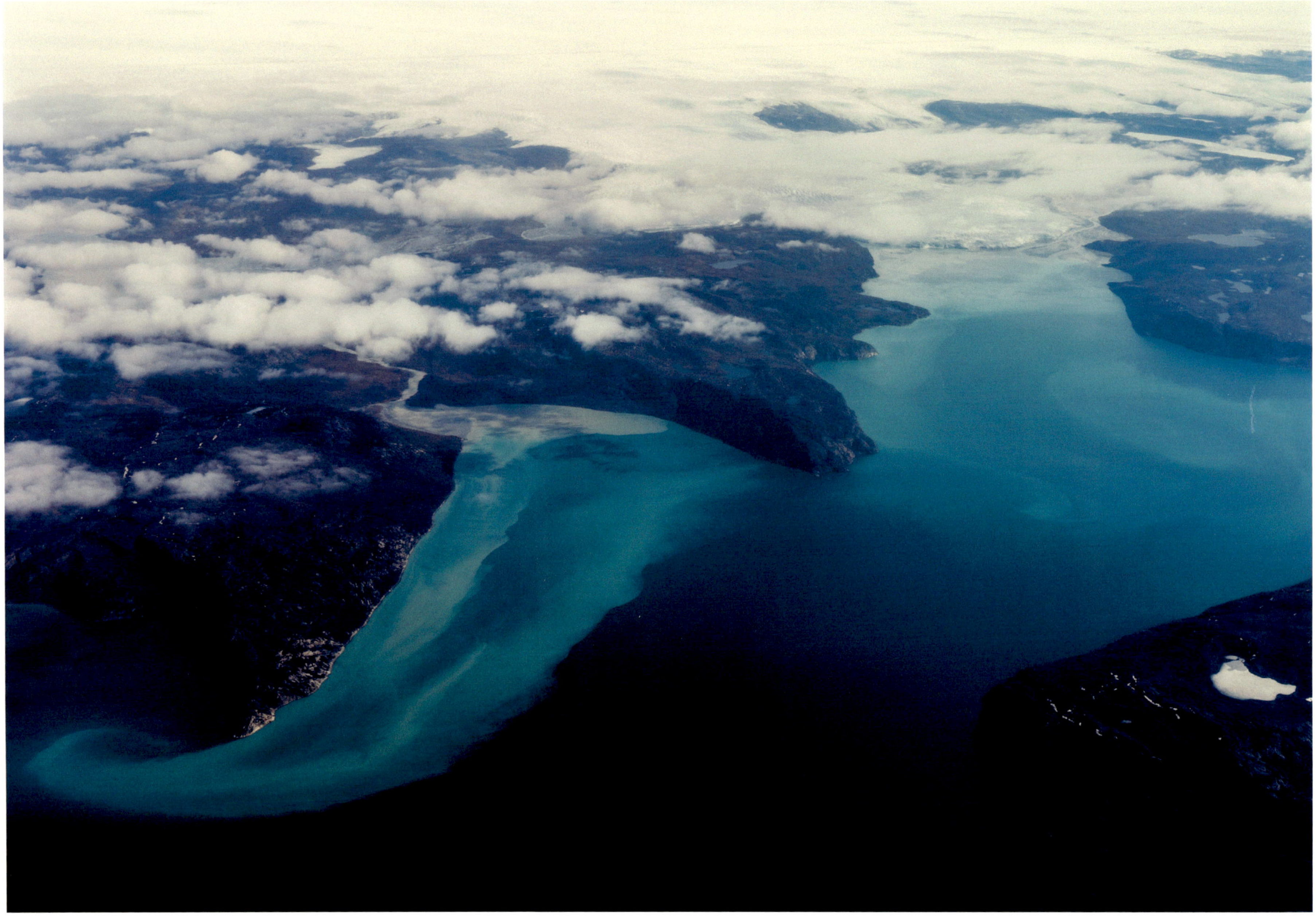

SISIMIUT AREA 2017
Aerial view of ice cap, fjord, and glacial run-off between Sisimiut and Nuuk.

# Nuuk to Sisimiut

DENIS DEFIBAUGH

**Danish missionary Hans Egede founded Nuuk in 1728 as the first town in Greenland, although Vikings inhabited the region 1,000 years ago and Nuuk's oldest settlements date back 4,200 years.**

Today, Nuuk is the largest town and the capital of *Kalaallit Nunaat*—the Inuit name for Greenland—with 17,316 residents.[1] Located on the southwestern coast, it is surrounded by the massive Nuup Kangerlua fjord system. A contemporary city with a strong Danish influence, Nuuk boasts fine restaurants, the National Museum, the Katuaq Cultural Center (its unique architecture inspired by waving Northern Lights), and *Ilisimatusarfik*—the University of Greenland—which was established in 1987.

"Nuuk is a fantastic, exciting place," says Ann Andreasen, director of the Uummannaq Polar Institute. "There are plenty of stories from there...it's nice to have such a multicultural environment."[2]

Today's Nuuk stands in sharp contrast to what Rockwell Kent encountered when he arrived in June 1929 on the cutter *Direction*. Kent's first view of a Greenland community, in fact, was the settlement of Narsaq, just south of Nuuk. While anchored in a nearby fjord, a storm blew through, sinking *Direction* on the rocks. The same morning, Kent set out across the barren landscape to find help, and thus began his love affair with Greenland.

Kent spent time in Nuuk that summer waiting for *Direction* to be repaired. As luck would have it, a few months later he met Knud Rasmussen, a Greenlandic-Danish explorer and anthropologist, on a ship from Nuuk to Copenhagen, and lived with him for a short time while completing illustrations for an edition of *Moby Dick*. With Rasmussen's help and encouragement, Kent began planning his return to Greenland for the summer of 1931; it was also Rasmussen who suggested Illorsuit as the location for Kent's new home.

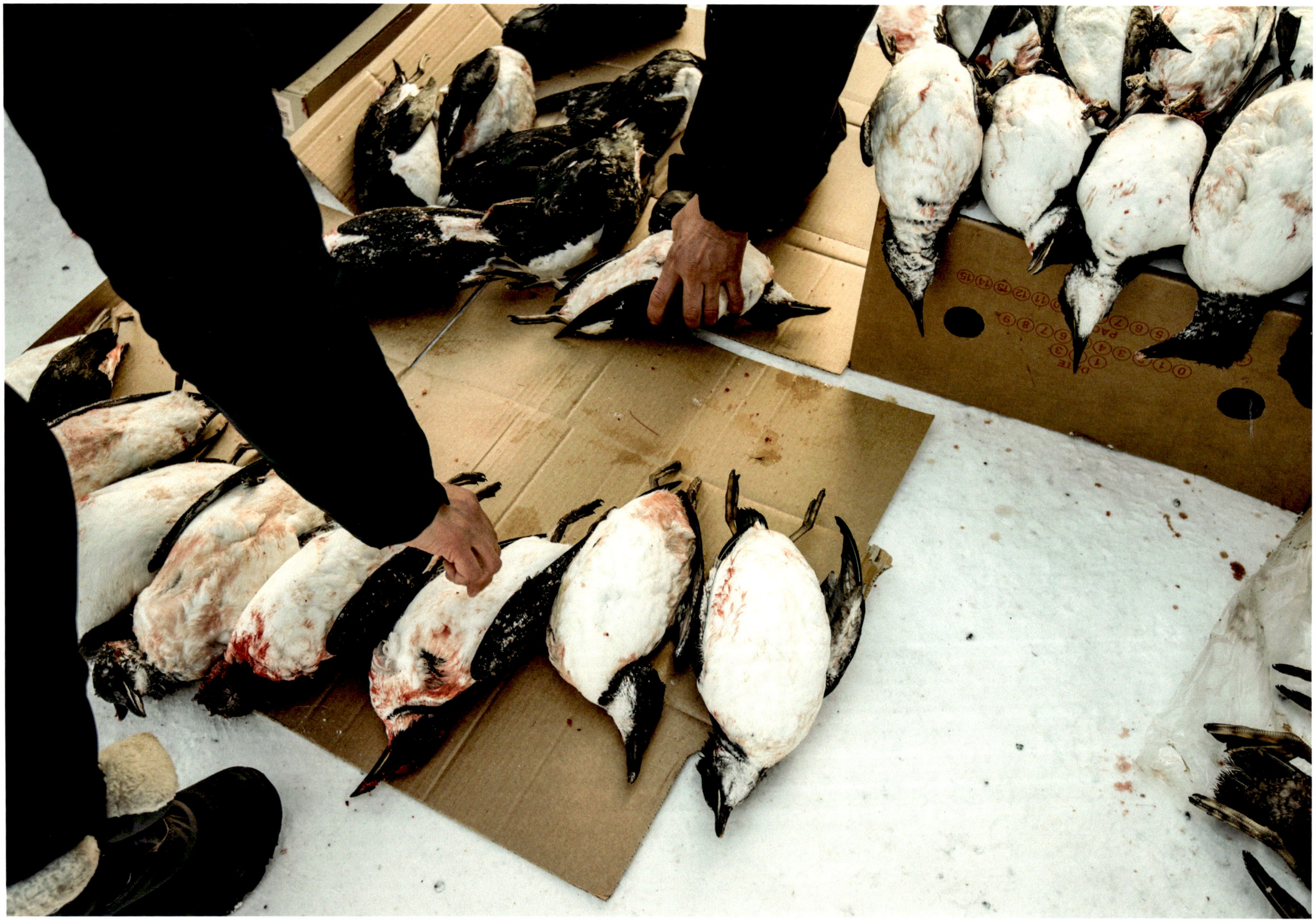

NUUK 2016
Hunters preparing to sell Appa,
the catch of the day.

Two years later, in 1932, Kent visited Sisimiut, about 150 miles north of Nuuk, while awaiting the arrival in Greenland of his wife, Frances. Sisimiut is Greenland's second largest city, with 5,539 residents in 2016.[3] In addition to its vibrant fishing community—the Royal Greenland company operates a large shrimp and cod processing factory—Sisimiut also is known for its independent minded and proud residents.

"Some joke that Sisimiut is actually Greenland's largest city, because Nuuk is just Danish,"[4] quips Olafur Rafnar Olafsson, a photographer and eskimologist who works as a destination manager for Arctic Circle Business in Sisimiut. Adds Andreasen, "People of Sisimiut are self-confident. It's nice to know where you come from."[5]

Although Kent's time in Sisimiut was short, it was productive. He made a substantial number of photographs, spending time in the town manager's home and photographing the community's people and landscapes. His portraits, usually of a single person posing for the camera near the manager's home, provide a descriptive narrative of both the town and its harbor and are yet another of Kent's many valuable contributions to documenting Greenlandic life in the 1930s.

**Notes**

1 Bolatta Yahl and Naduk Kleemann, eds., *Statistics Greenland*, 13th ed. (Greenland: Statistics of Greenland, 2016), 9.

2 Ann Andreasen (director, Uummannaq Polar Institute), in discussion with Jette Rygaard, September 2016.

3 *Statistics Greenland*, 9.

4 Olafur Rafnar Olafsson (destination manager, Arctic Circle Business) interview with the author, November 2016.

5 Andreasen.

SISIMIUT 2017
Niviana Endesen and friends playing jump rope in the middle of winter.

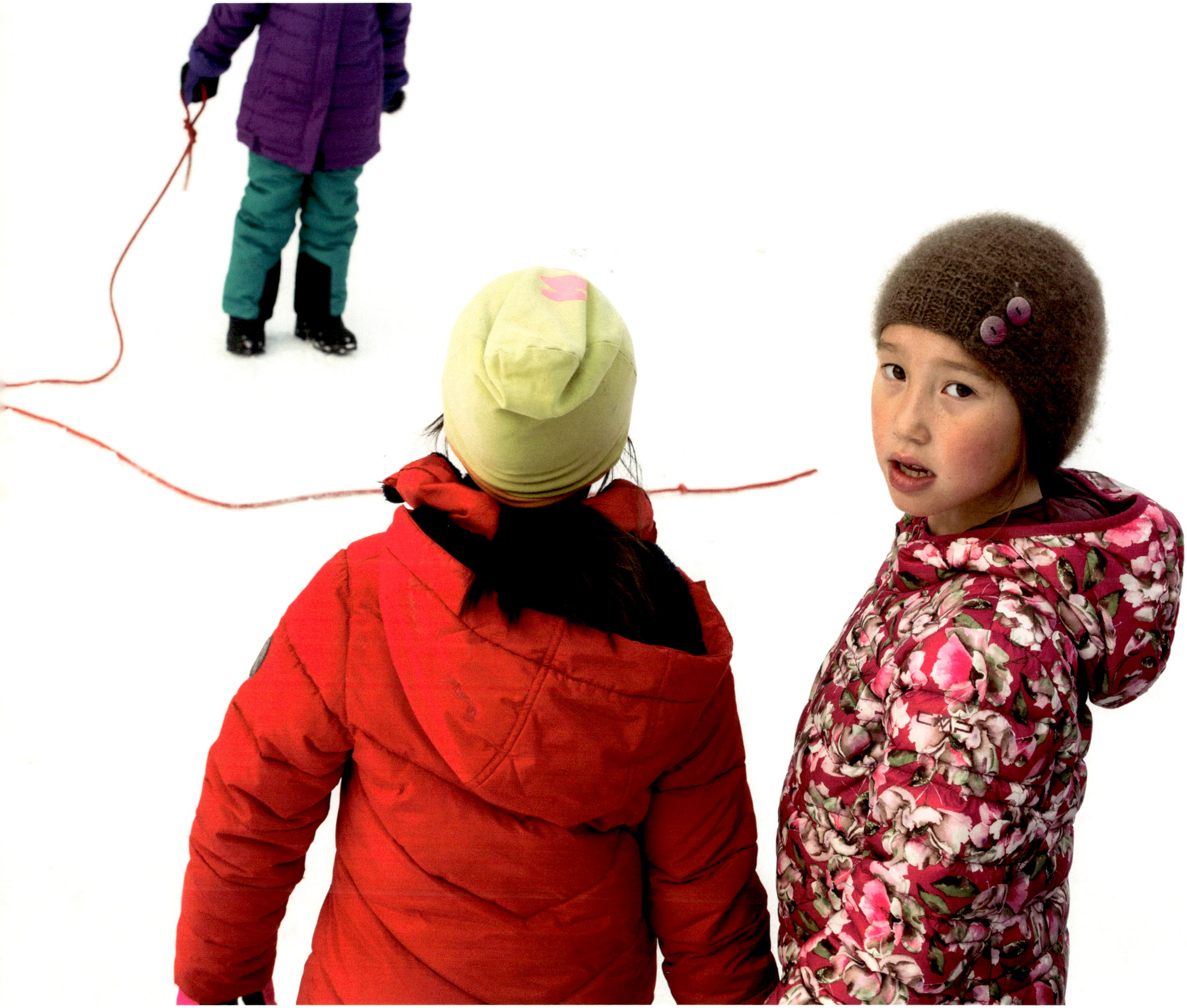

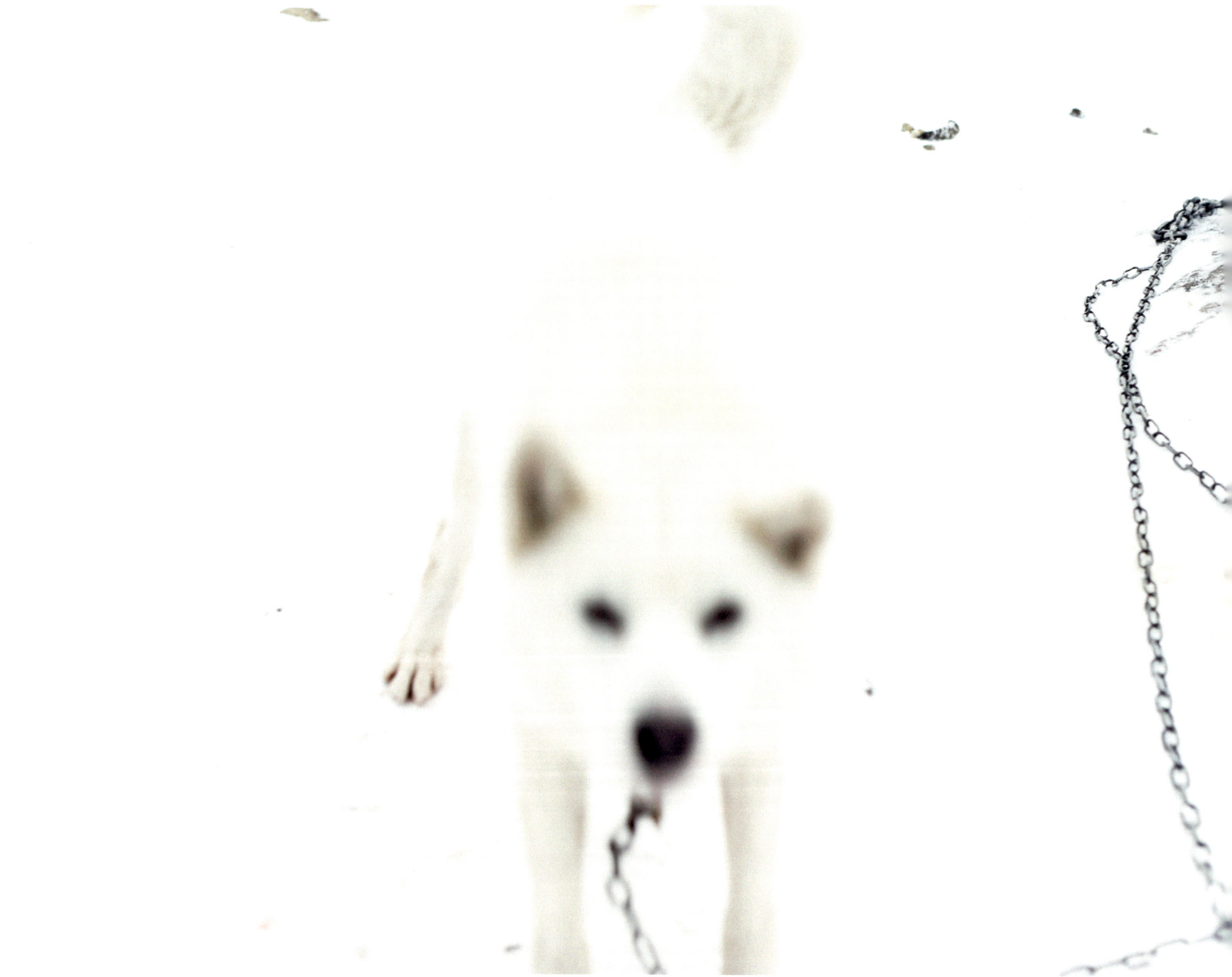

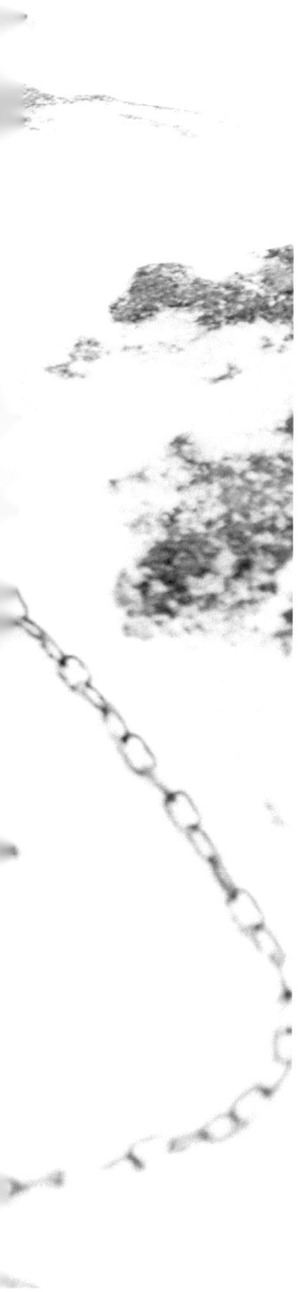

SISIMIUT 2017
Dogs are chained at six months old and remain so unless they are pulling a sledge.

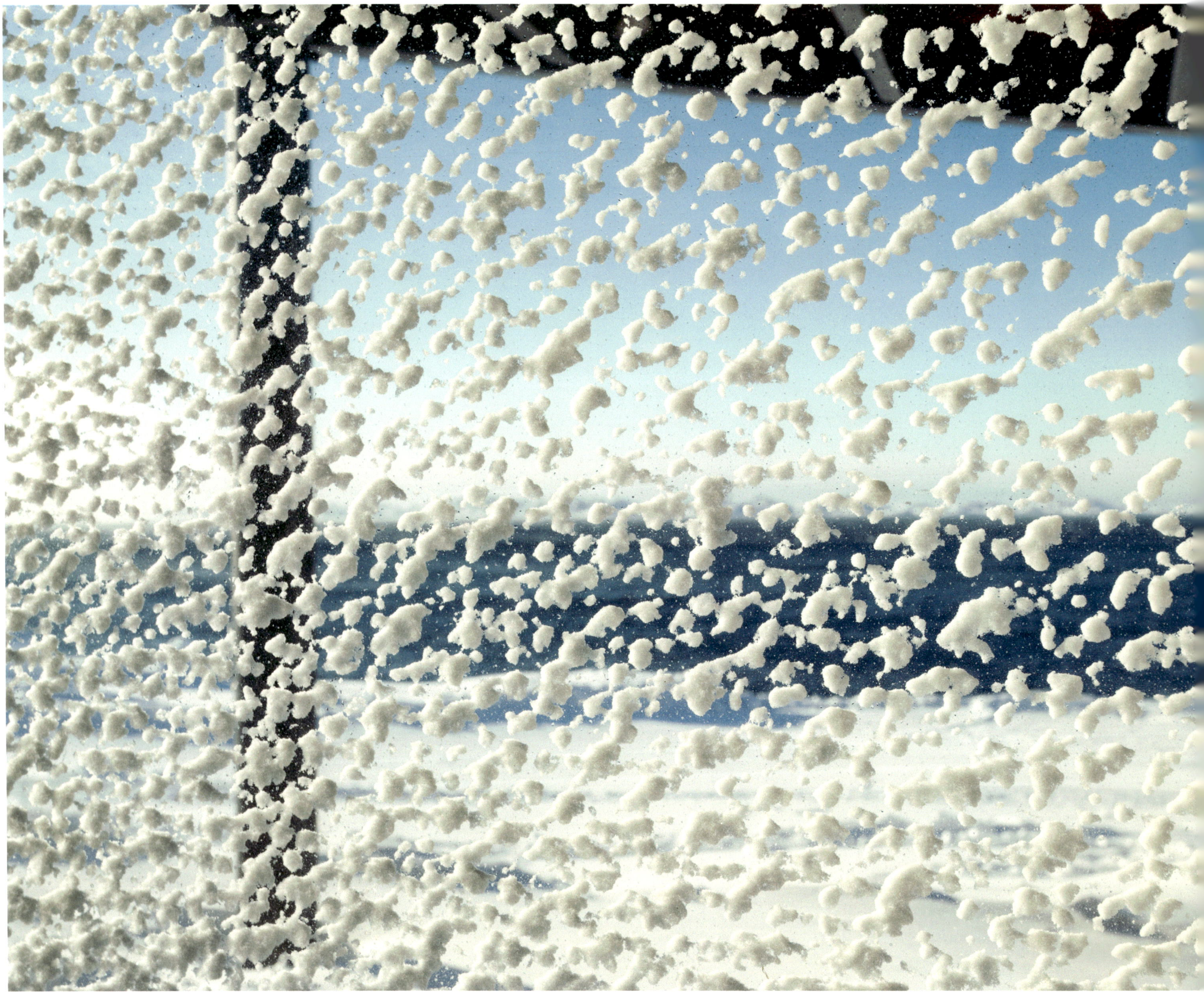

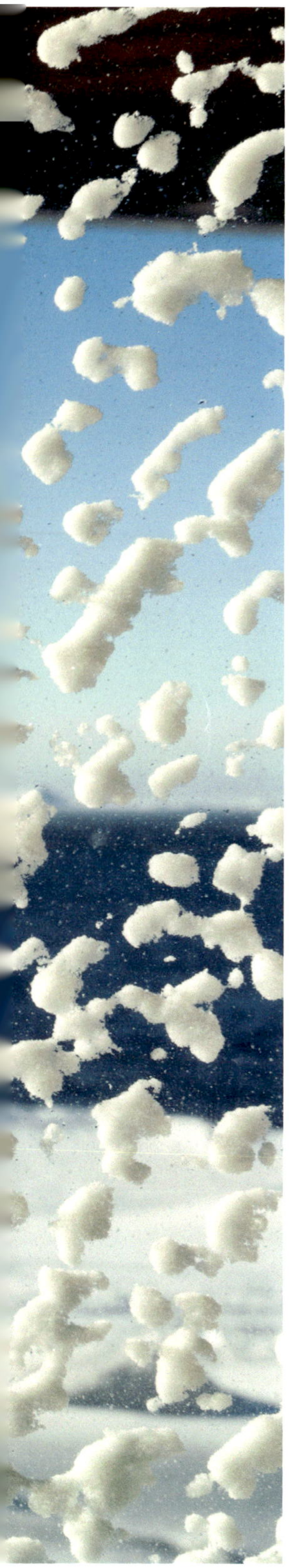

NUUK 2017

View from window following a snowstorm.

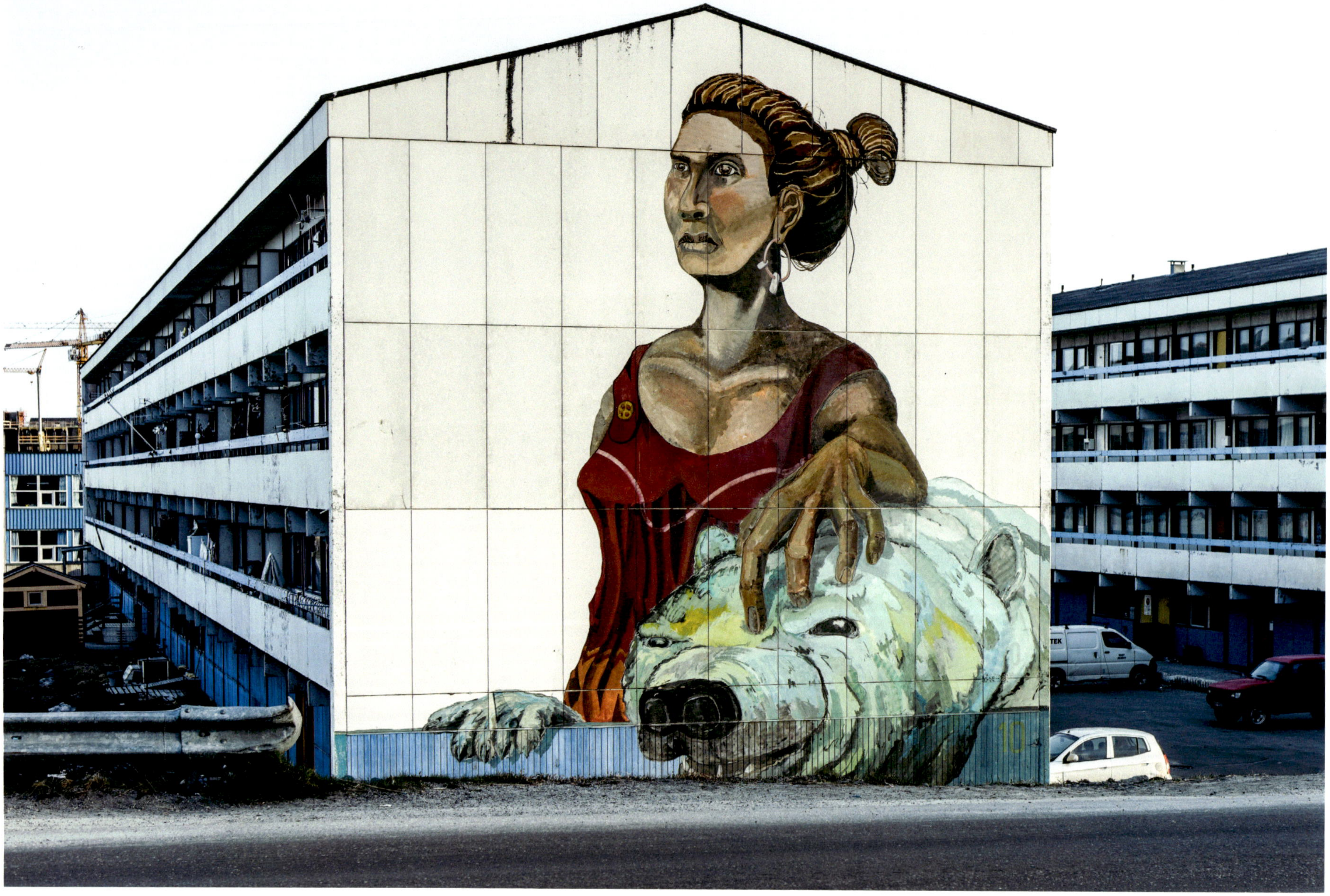

NUUK 2017

A mural of an Inuit woman and bear highlights one of the many lower-income blok house buildings.

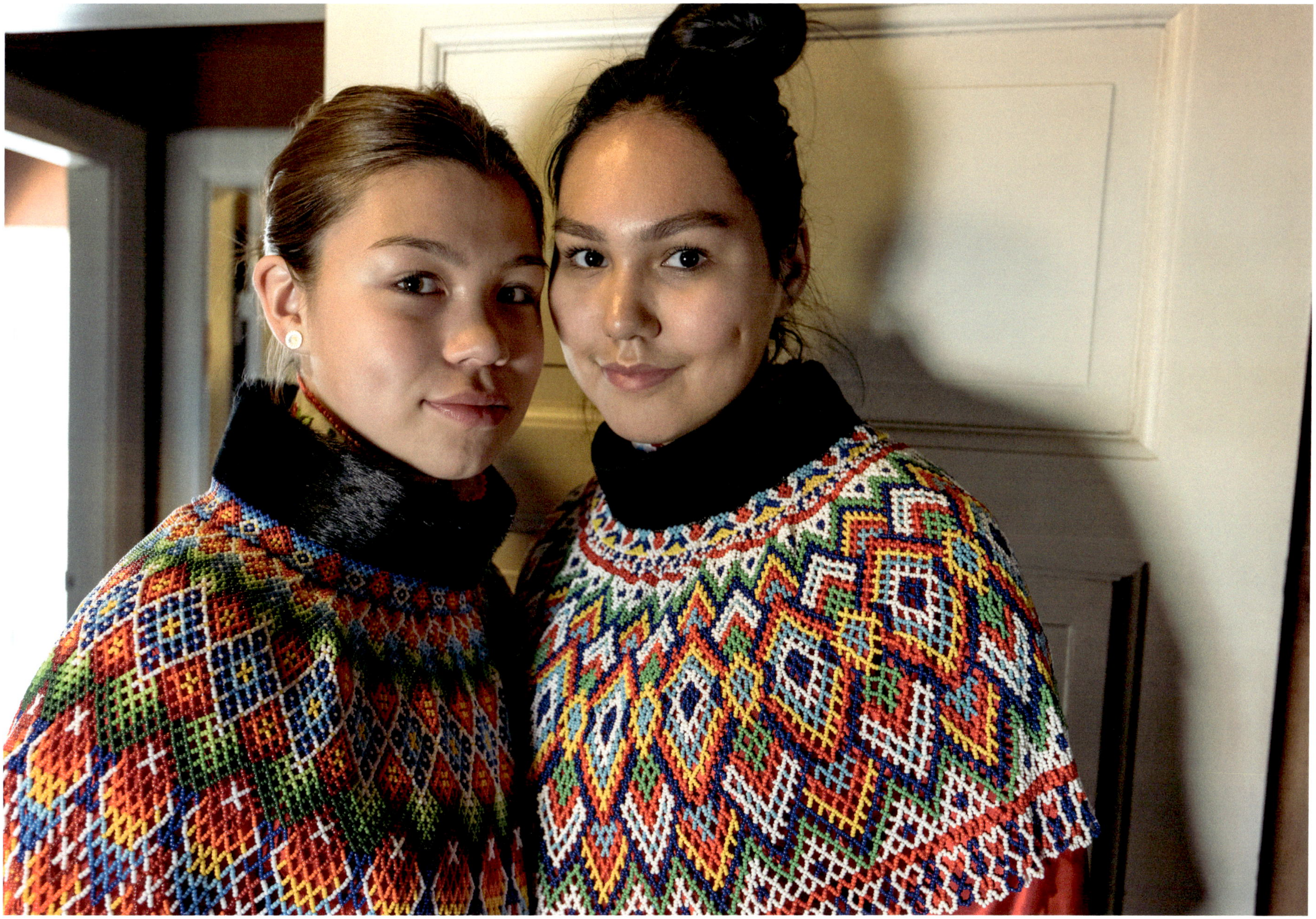

SISIMIUT 2017
Students Ajaaja Silasen and Maria Lennert attending Confirmation in traditional yoke-sewn sweaters with thousands of glass beads.

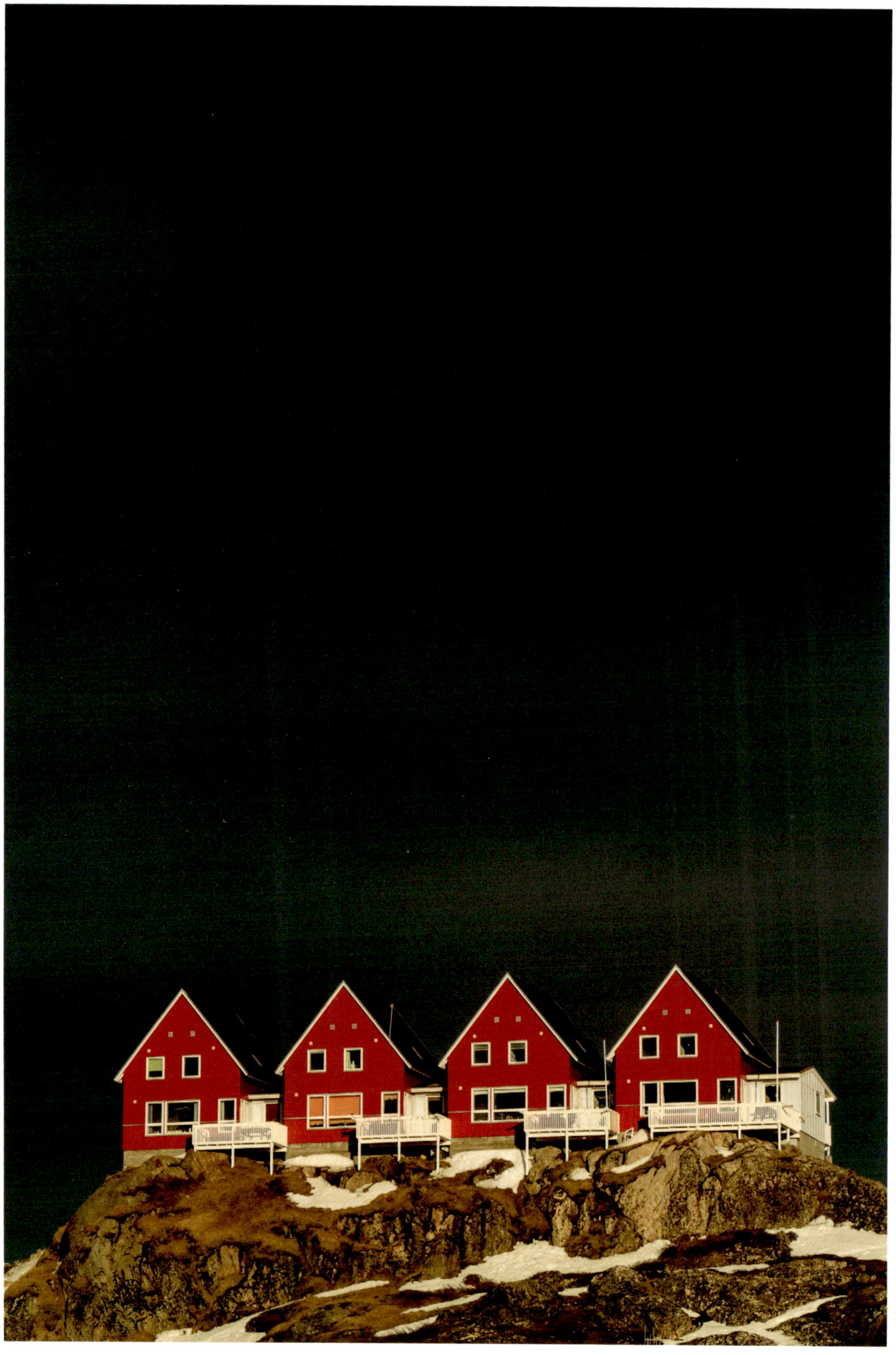

SISIMIUT 2017

Dramatic environmental architecture.

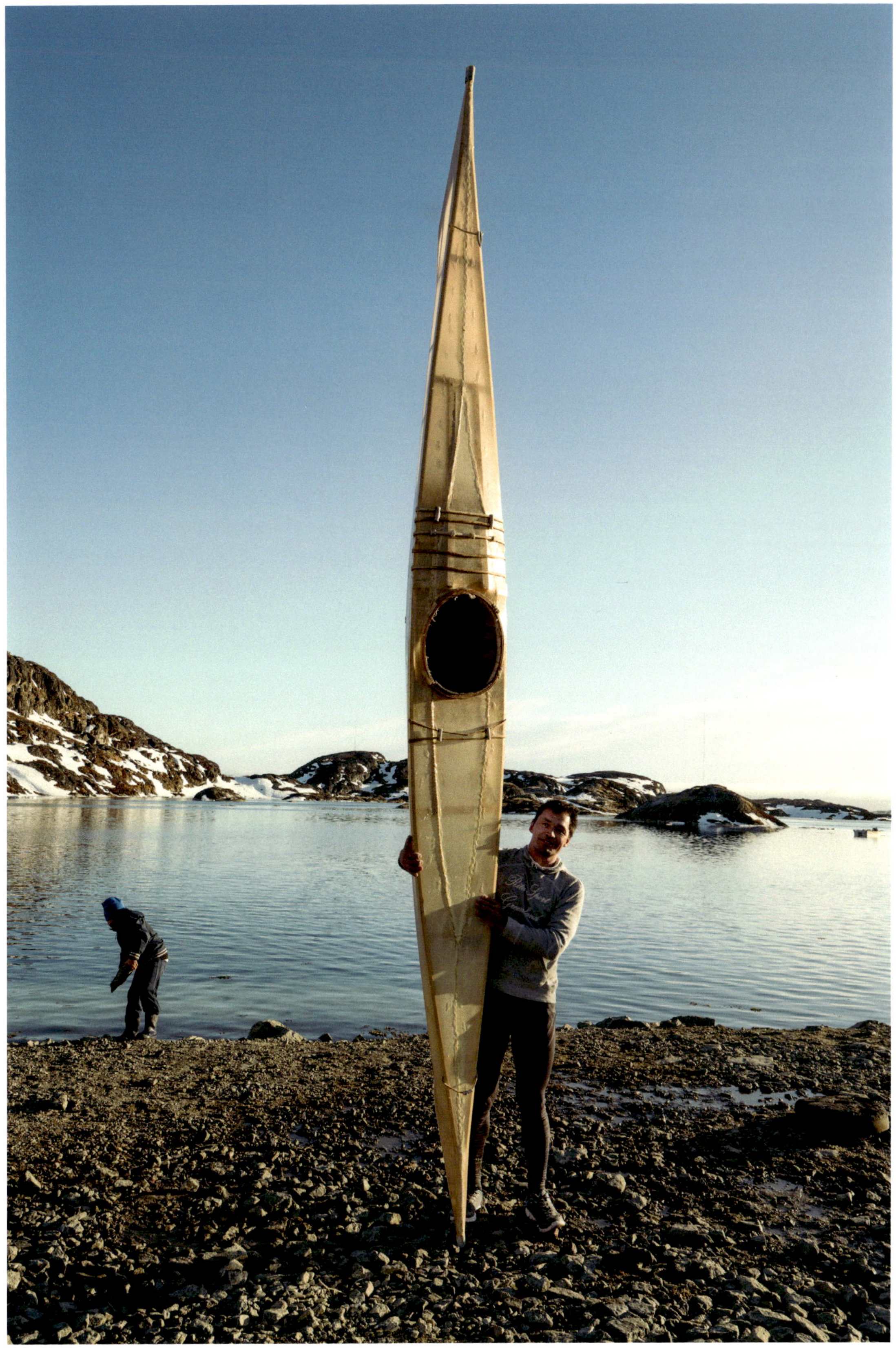

SISIMIUT 2017
Kayak builder Jorgen Leander introduced Tyvek® high-density polyethylene as an effective kayak skin material.

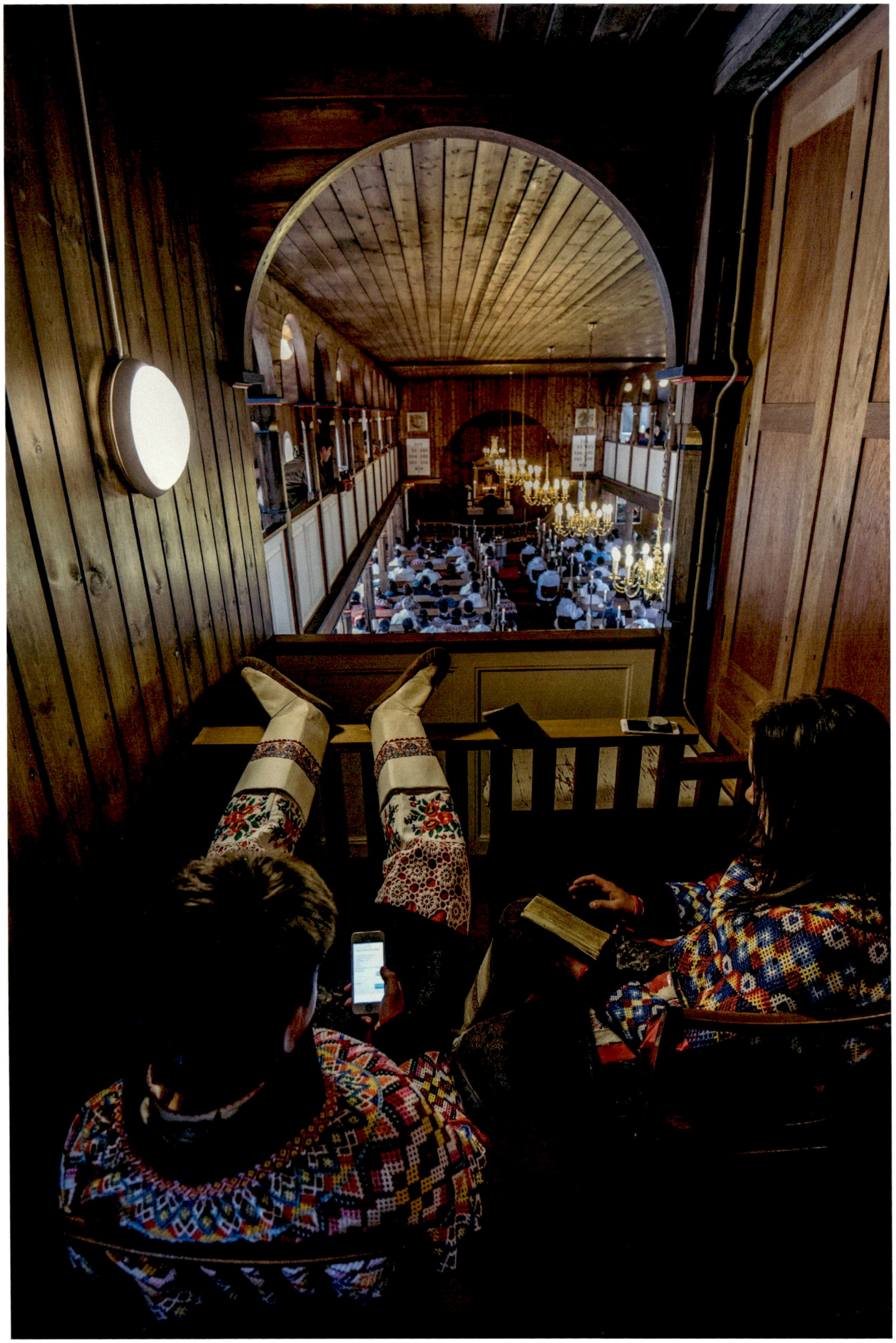

SISIMIUT 2017

The historic Sisimiut church built in 1926 celebrates the sacrament of Confirmation. Such ceremonies can be long.

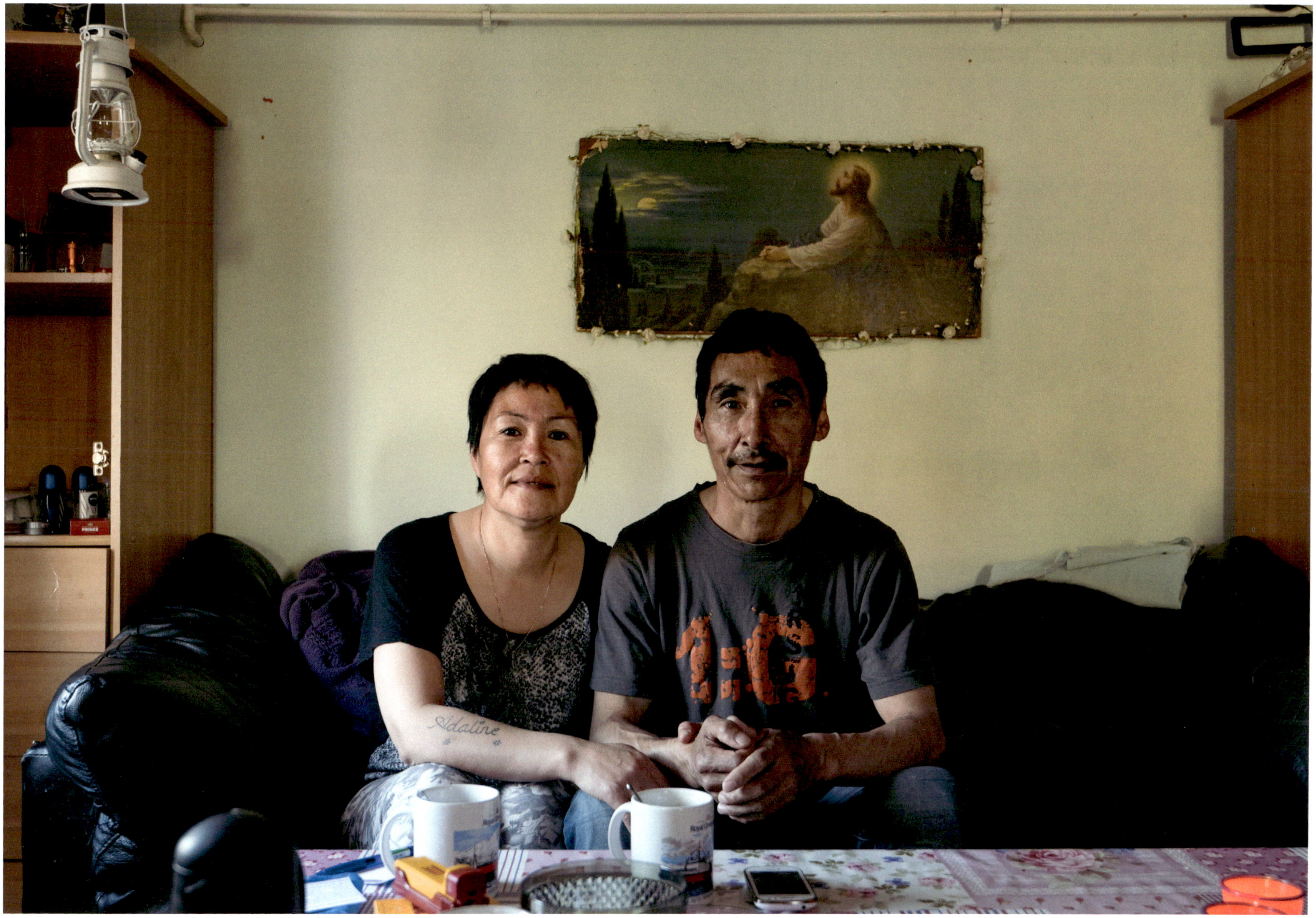


SISIMIUT 2016
Adaline Frederiksen and Neils Kliest
in their home.

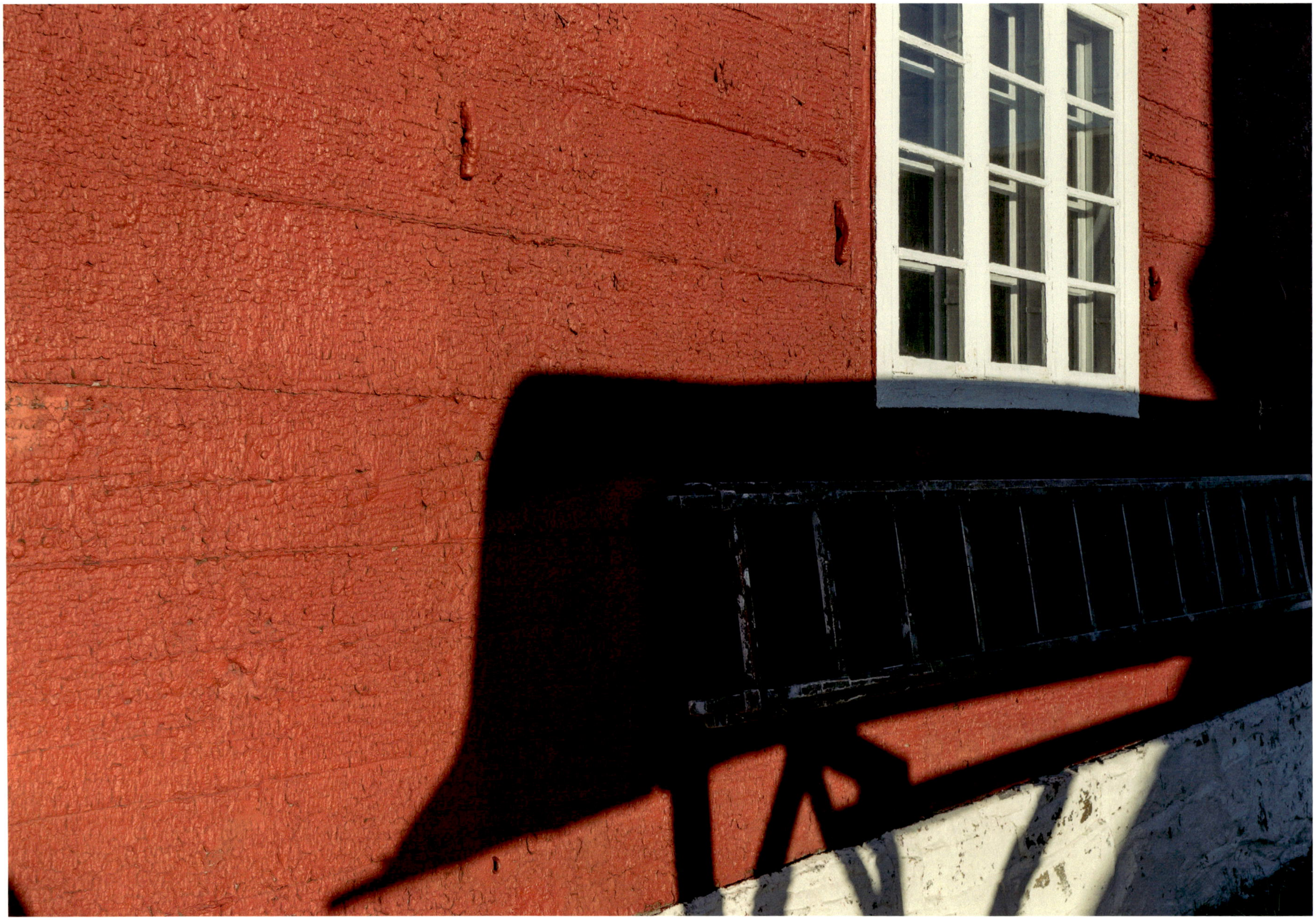

SISIMIUT 2016

Shadow of an *umiak*, or skin boat, typically used to transport family goods. These boats were often called “the woman’s boats,” as women did the rowing and moving.

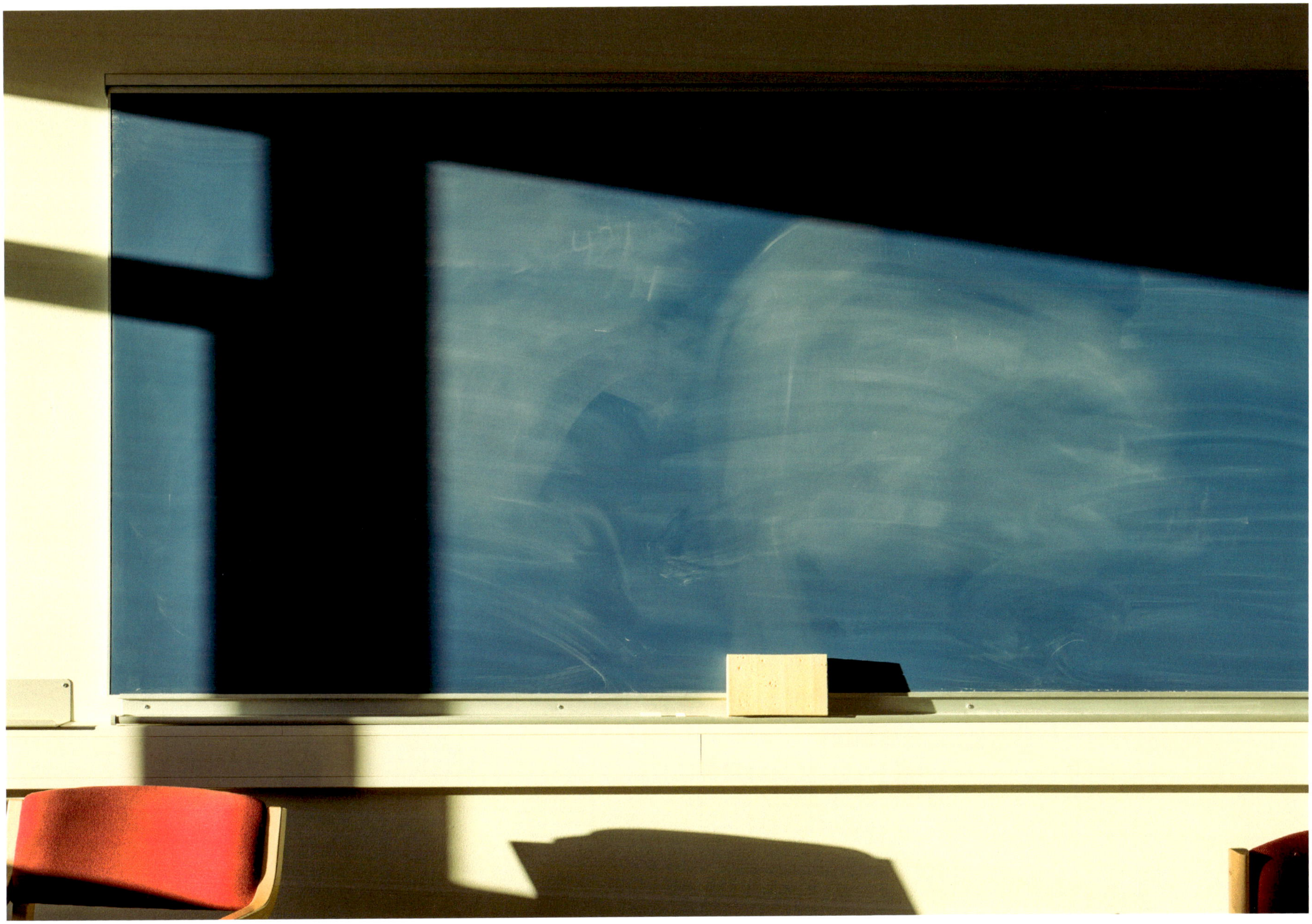

SISIMIUT 2016
A typical Sisimiut classroom chalkboard as seen following a photography workshop session with Inuit students.

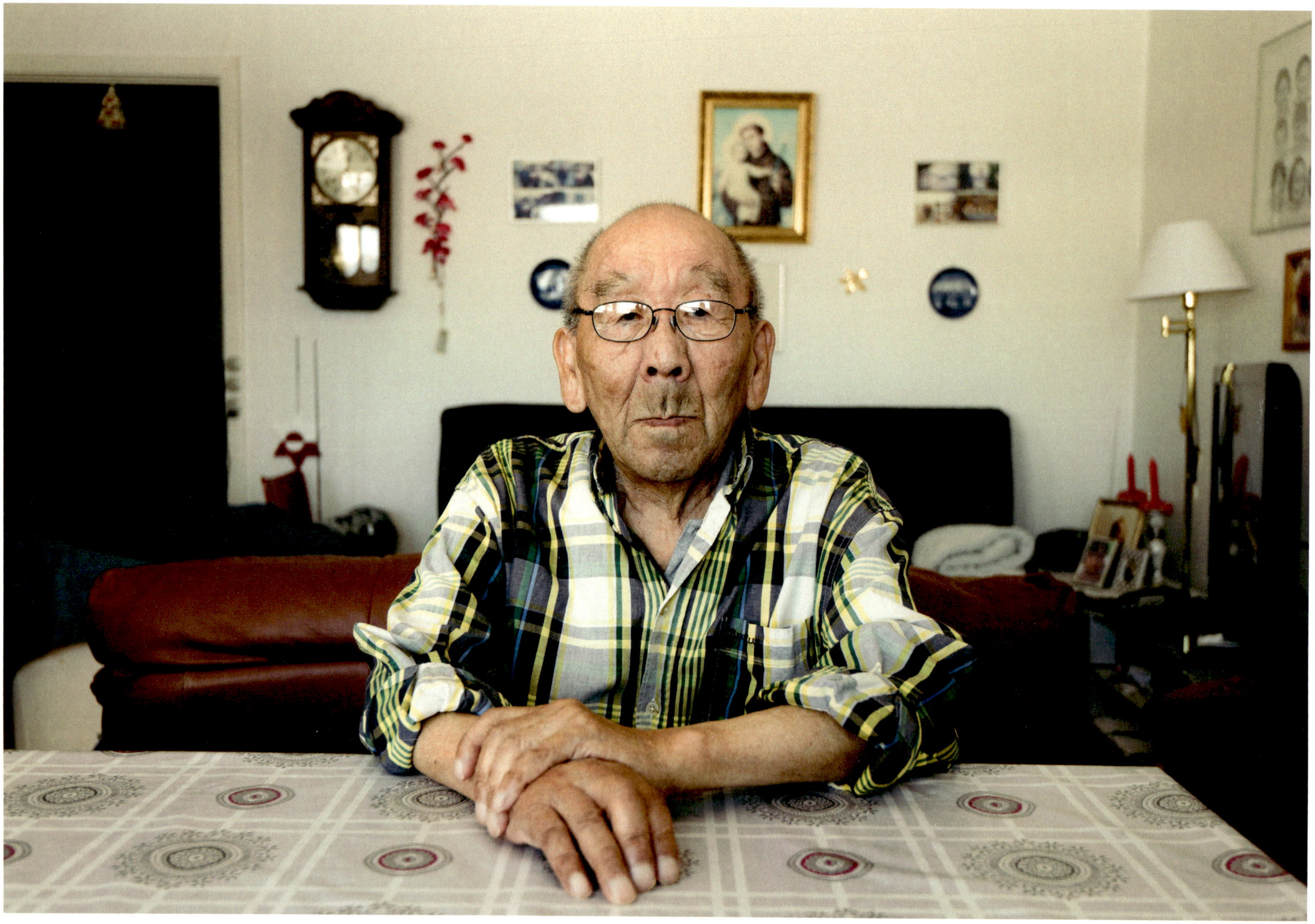

NUUK 2017
Amos Egede, a radio personality and comedian, at home.

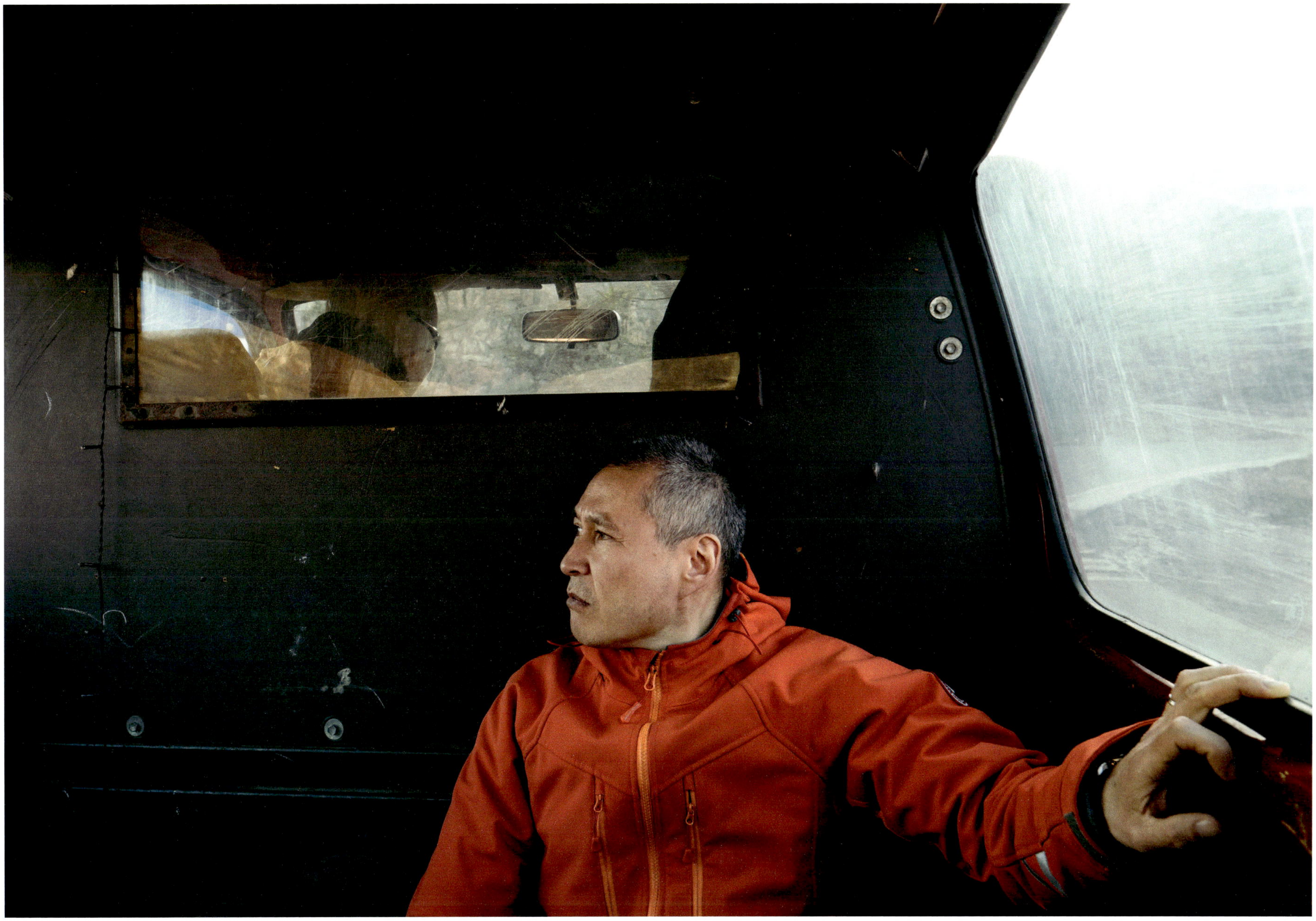

SISIMIUT 2016
Axel Jeremiassen and an unnamed Discovery Channel journalist travel to test a drone.

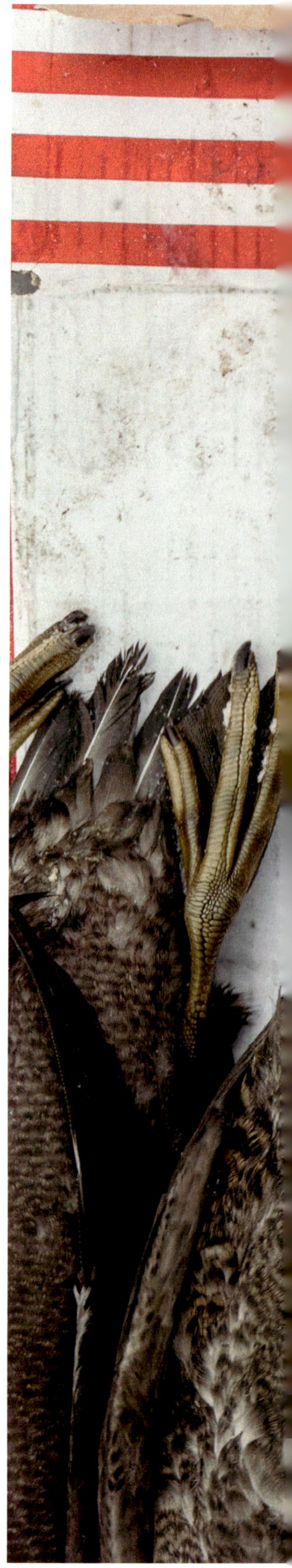

SISIMIUT 2016

Freshly killed ducks for sale are displayed on a cardboard box for “Real American” hamburger buns.

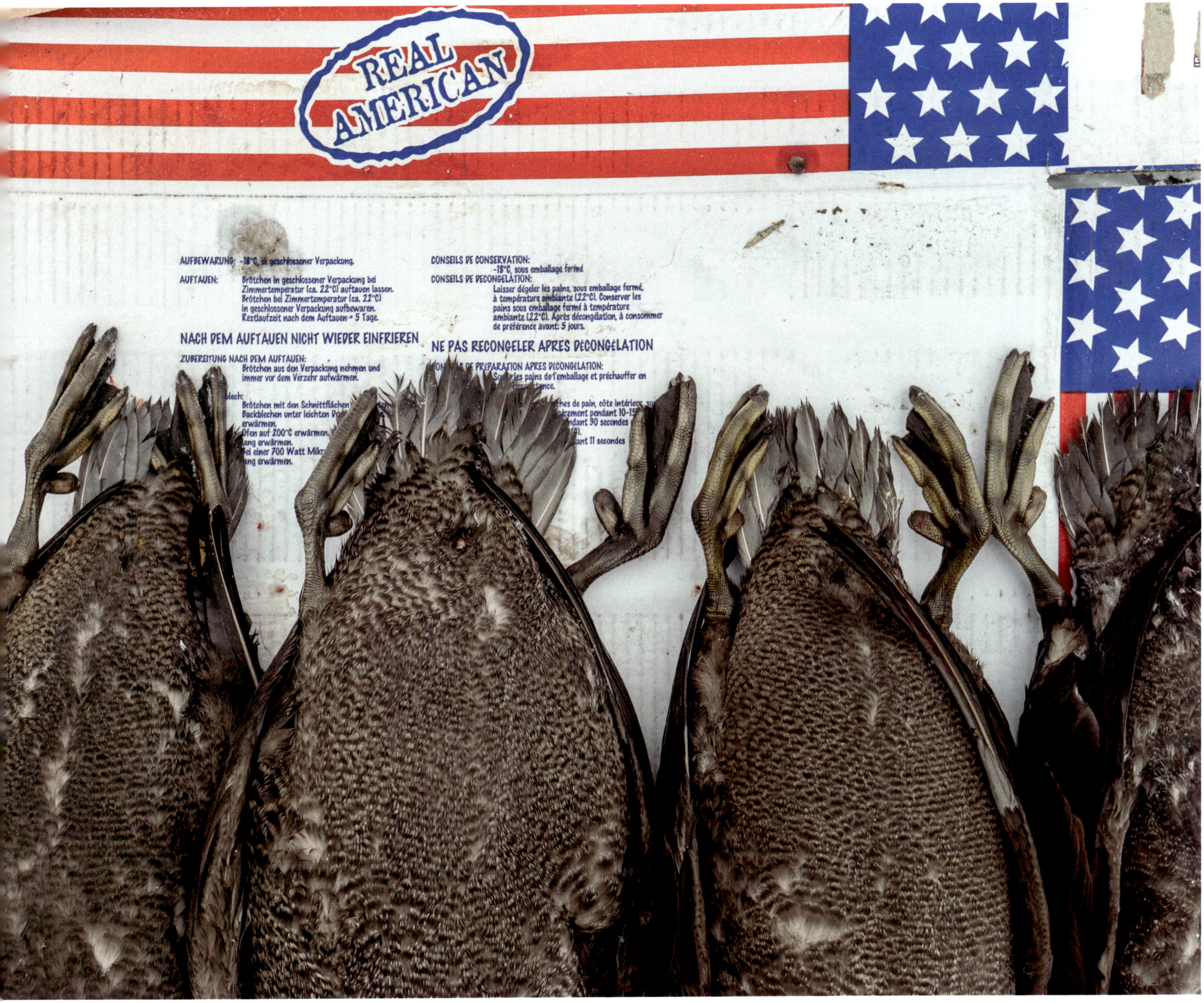
REAL AMERICAN
AUFBEWARUNG: -18°C in geschlossener Verpackung.
AUFTAUEN: Brötchen in geschlossener Verpackung bei Zimmertemperatur (ca. 22°C) auftauen lassen. Brötchen bei Zimmertemperatur (ca. 22°C) in geschlossener Verpackung aufbewaren. Restlaufzeit nach dem Auftauen = 5 Tage.
NACH DEM AUFTAUEN NICHT WIEDER EINFRIEREN
ZUBEREITUNG NACH DEM AUFTAUEN: Brötchen aus den Verpackung nehmen und immer vor dem Verzehr aufwärmen.
CONSEILS DE CONSERVATION: -18°C, sous emballage fermé
CONSEILS DE DECONGELATION: Laisser dégeler les pains, sous emballage fermé, à température ambiante (22°C). Conserver les pains sous emballage fermé à température ambiante (22°C). Après décongélation, à consommer de préférence avant: 5 jours.
NE PAS RECONGELER APRES DéCONGéLATION

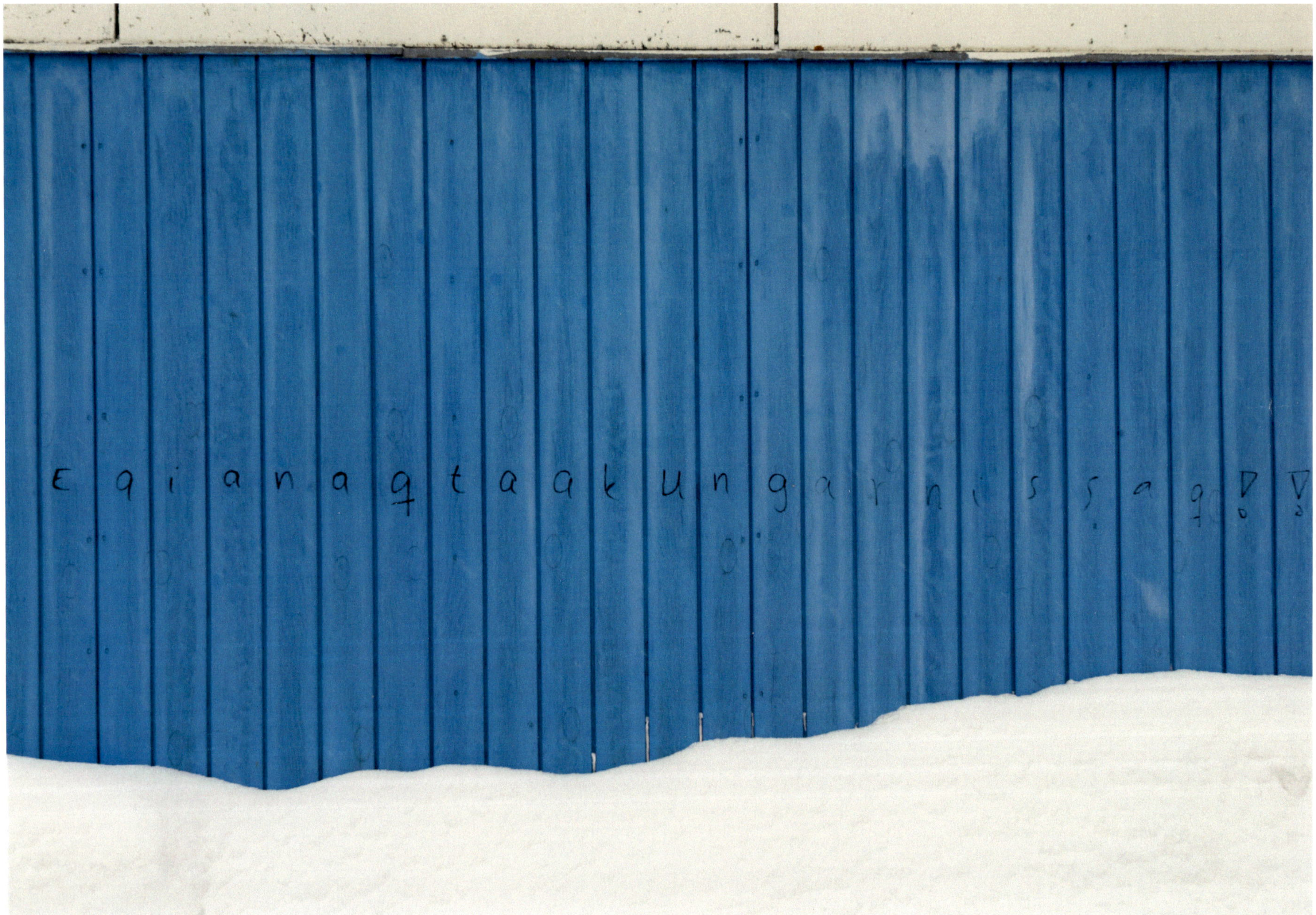


NUUK 2017
Graffiti on the side of a blok house. The rough translation: "I am too lazy to get there."

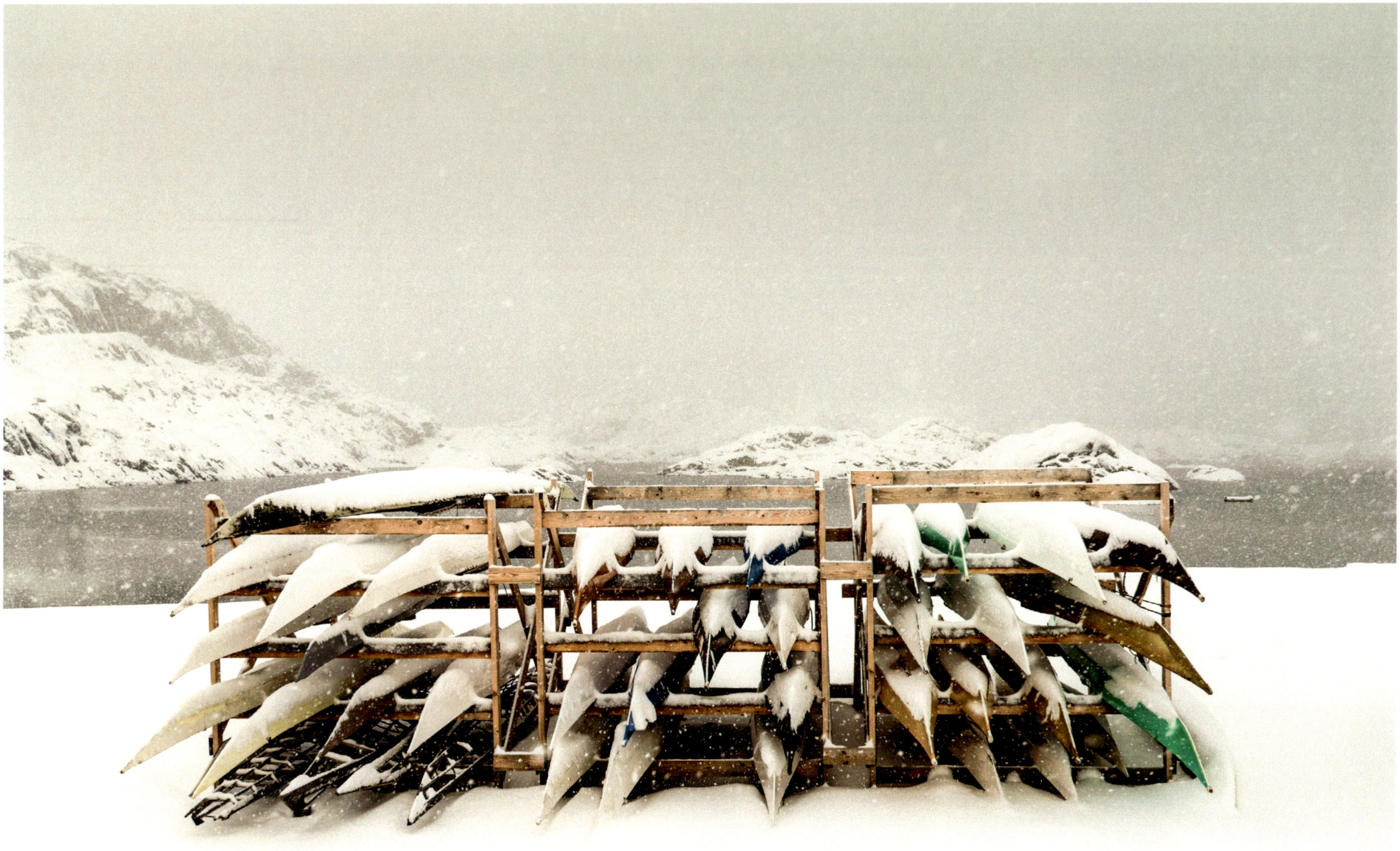

SISIMIUT 2017
Kayak rack in winter.

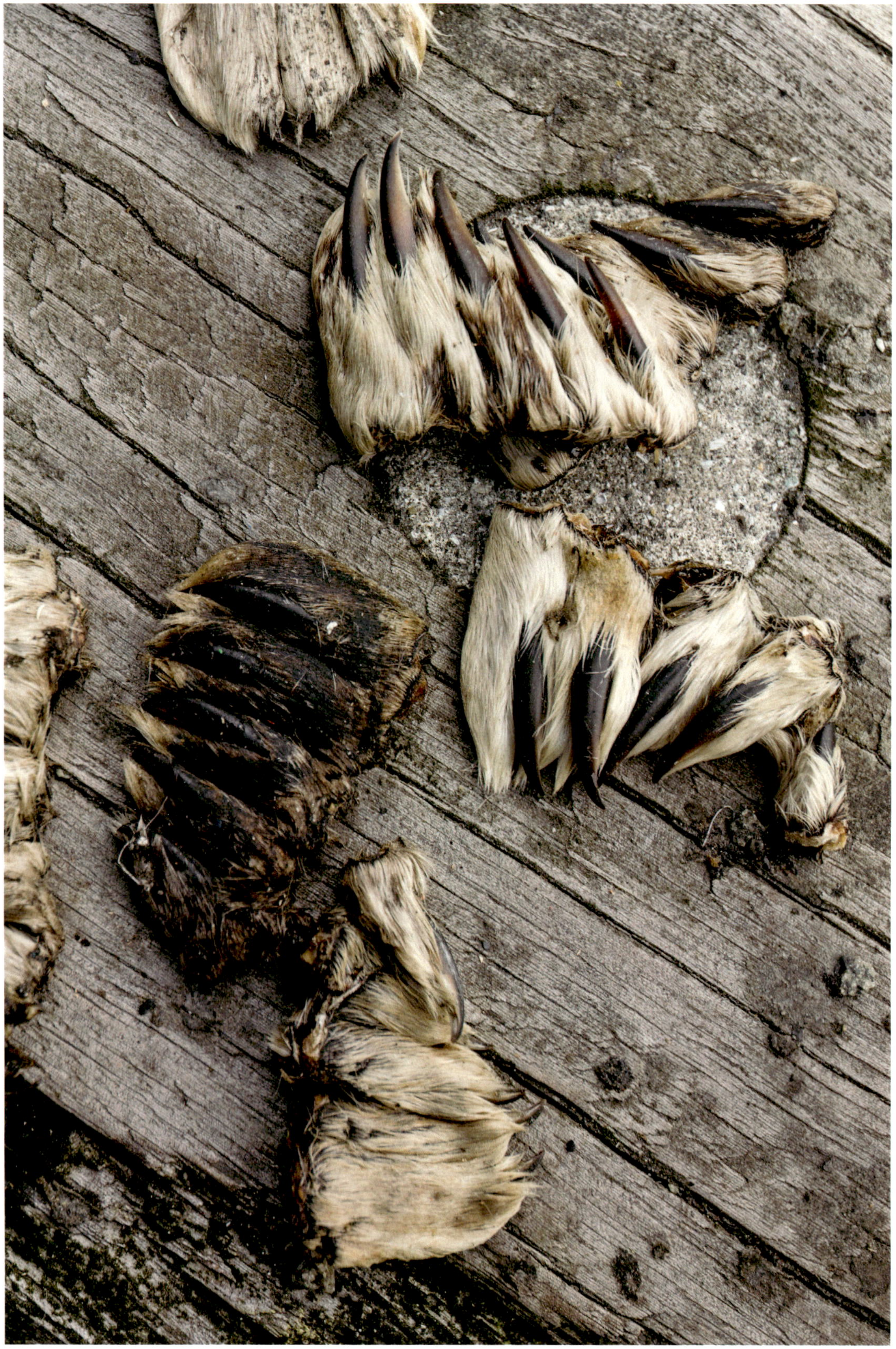

SISIMIUT 2016
Seal claws.

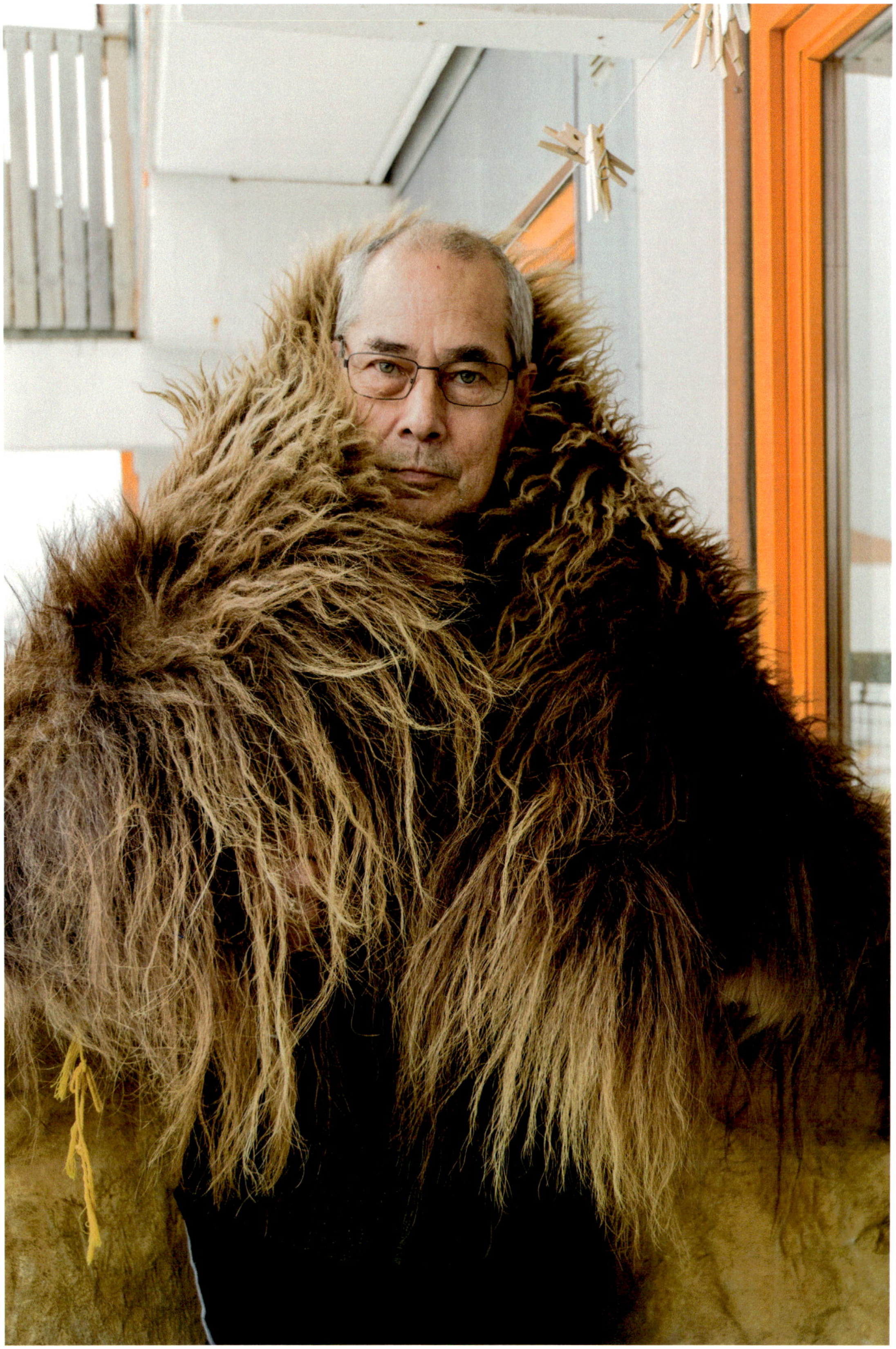

SISIMIUT 2016
Edward Qvist wearing muskox fur.

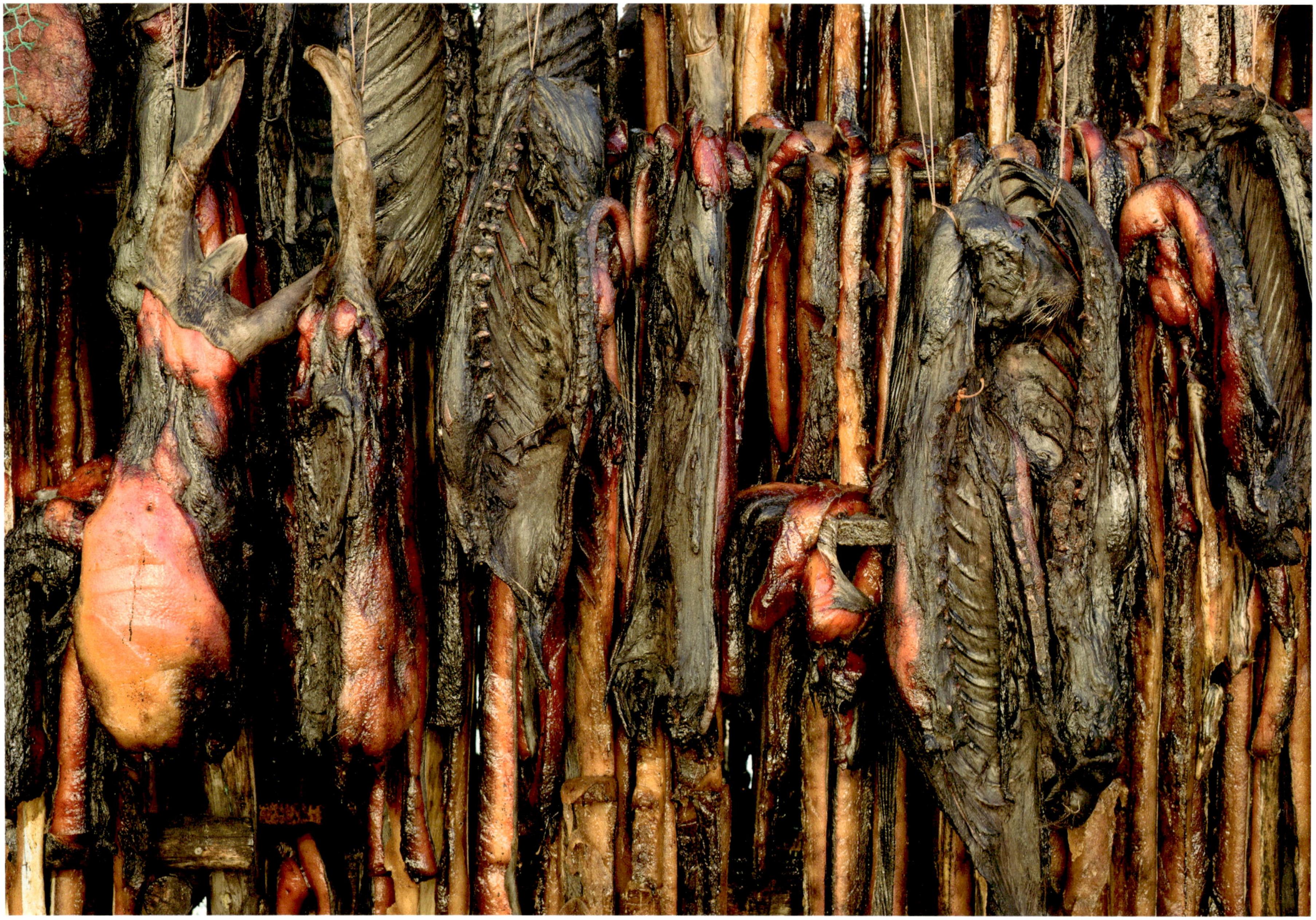

SISIMIUT 2016
Large drying rack of seal skins, ribs, and heads.

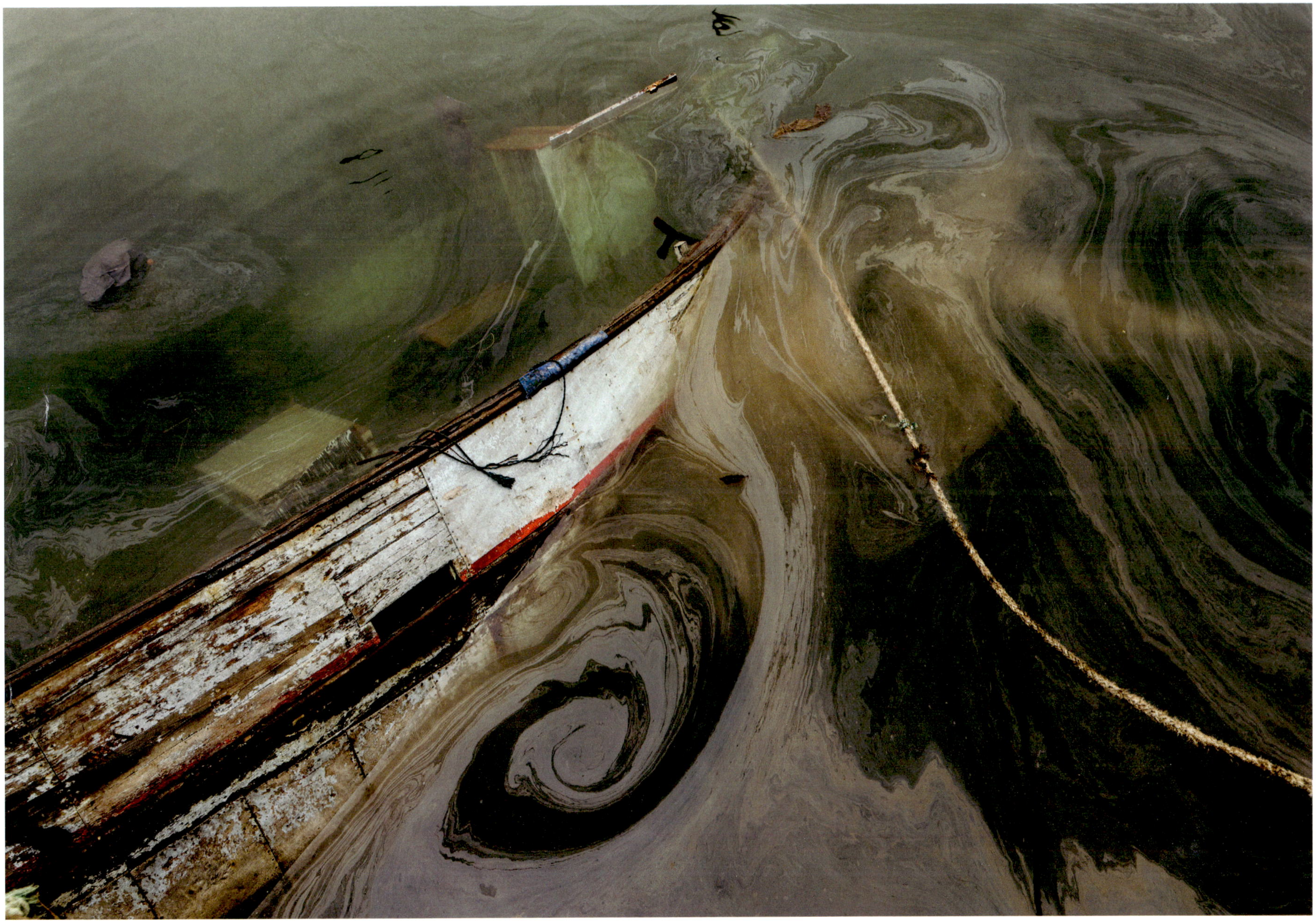

SISIMIUT 2016
A sunken boat leaks gasoline and oil into the harbor.

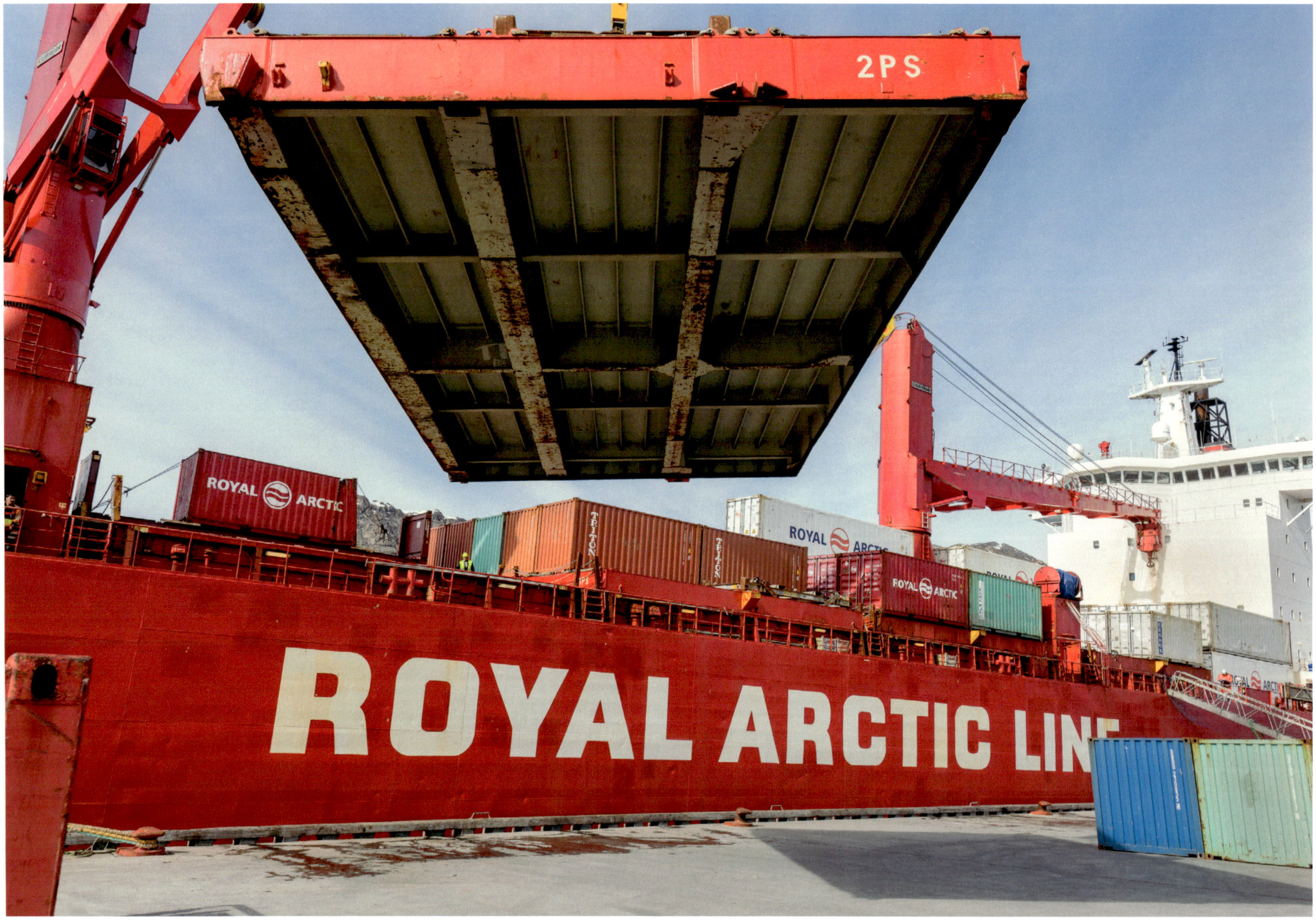


SISIMIUT 2017

A cargo ship unloads shipping containers
and supplies for West Greenland.

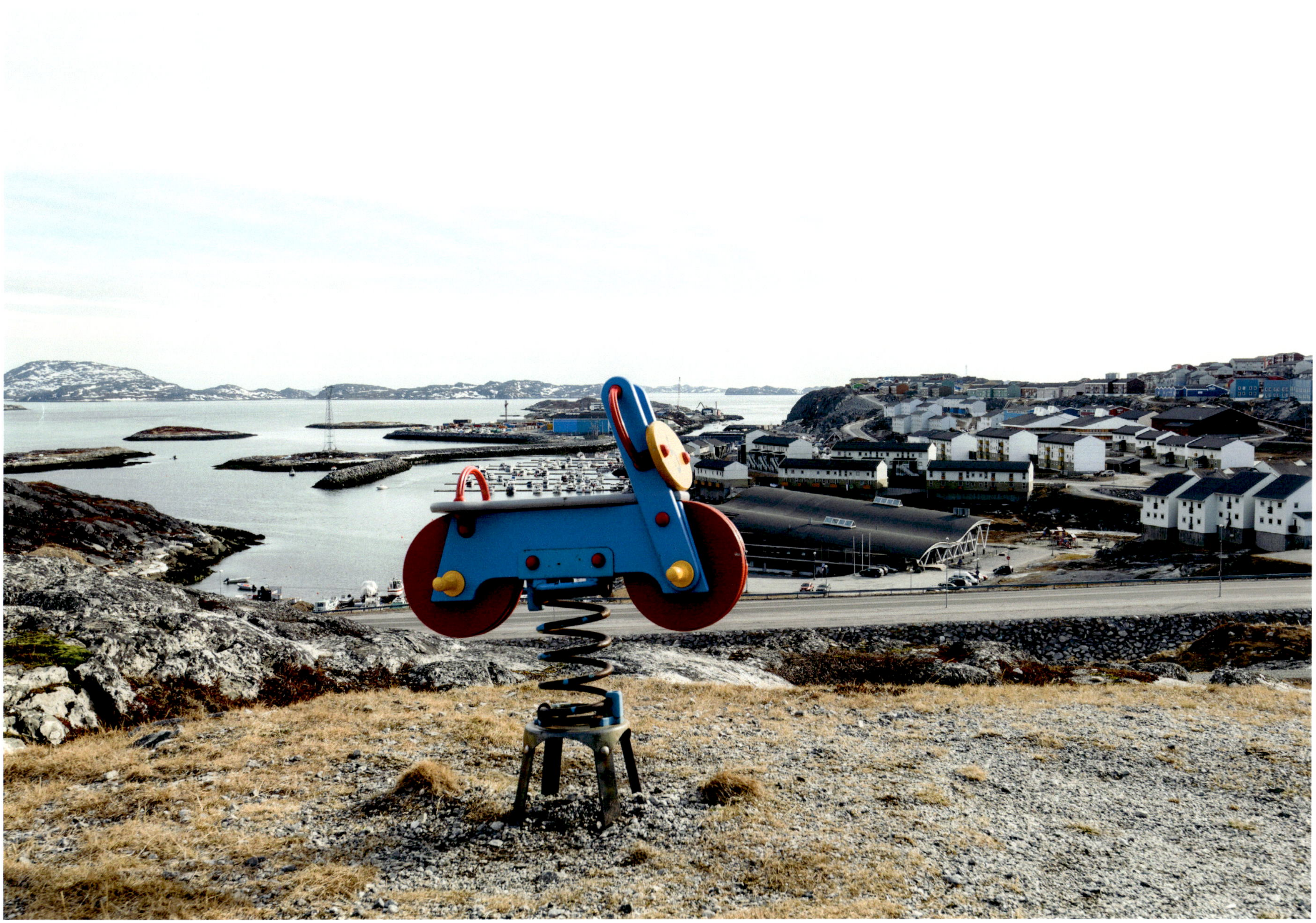

NUUK 2016
Playground with a view of
the swim center and harbor.

"WHEN RAVENS COULD SPEAK" from *INUIT FOLK TALES*

Once, long ago, there was a time when the ravens could talk. But strange thing about the raven's speech was that their words had the opposite meaning. When they wanted to thank any one, they used words of abuse, and thus always said the reverse of what they meant.

But as they were thus so full of lies, there came one day an old man, and by magic means took away their power of speech. And since that time the ravens can do no more than shriek.

But the raven's nature has not changed, and to this day they are an ill-tempered, lying, thieving lot.

Knud Rasmussen collection, *Inuit Folk Tales* (Hanover and London: IPI Press/University Press of New England, 2009), 128.

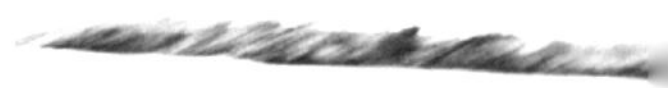

NUUK 2017
Ravens leaving the roost.

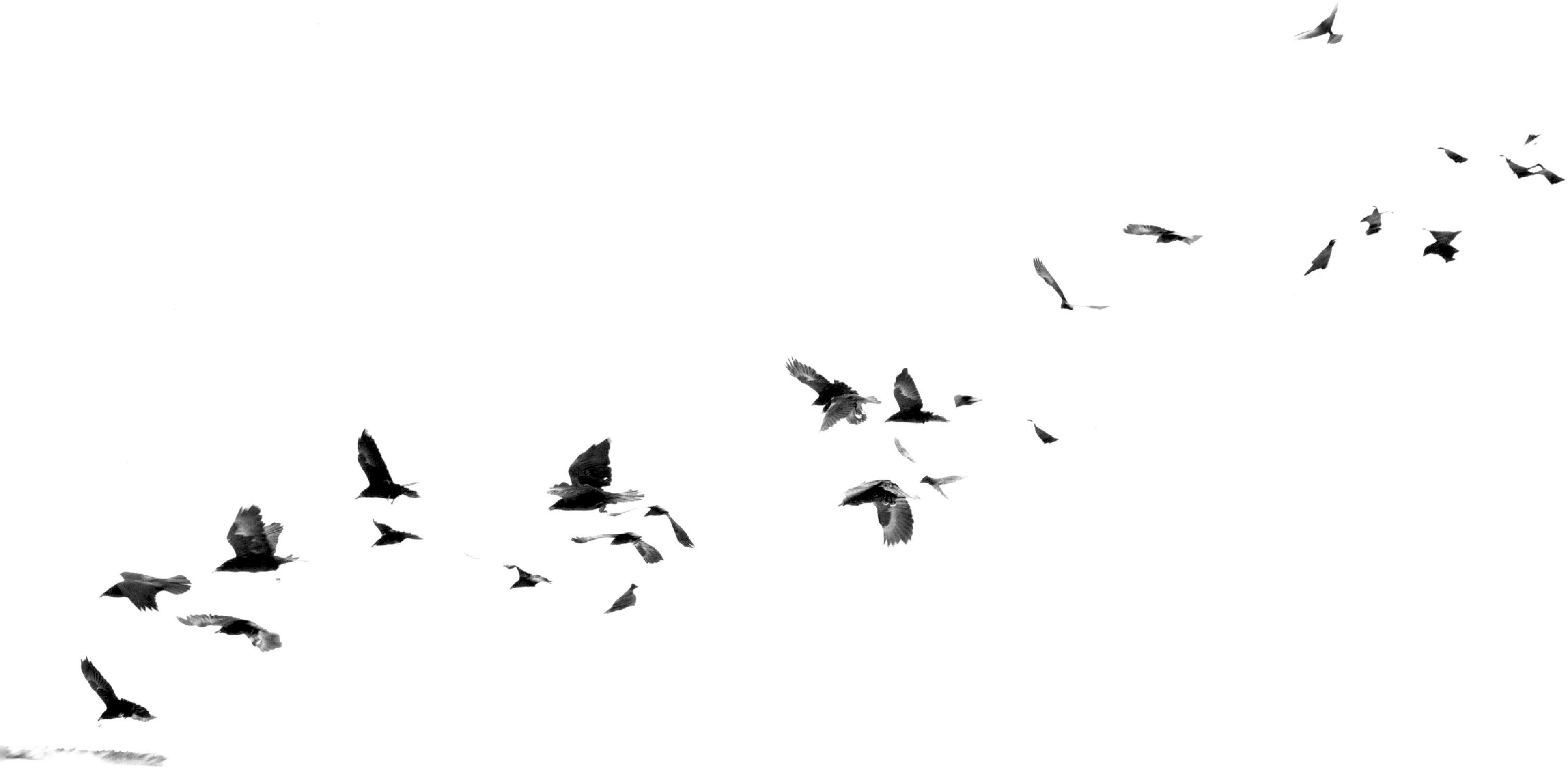

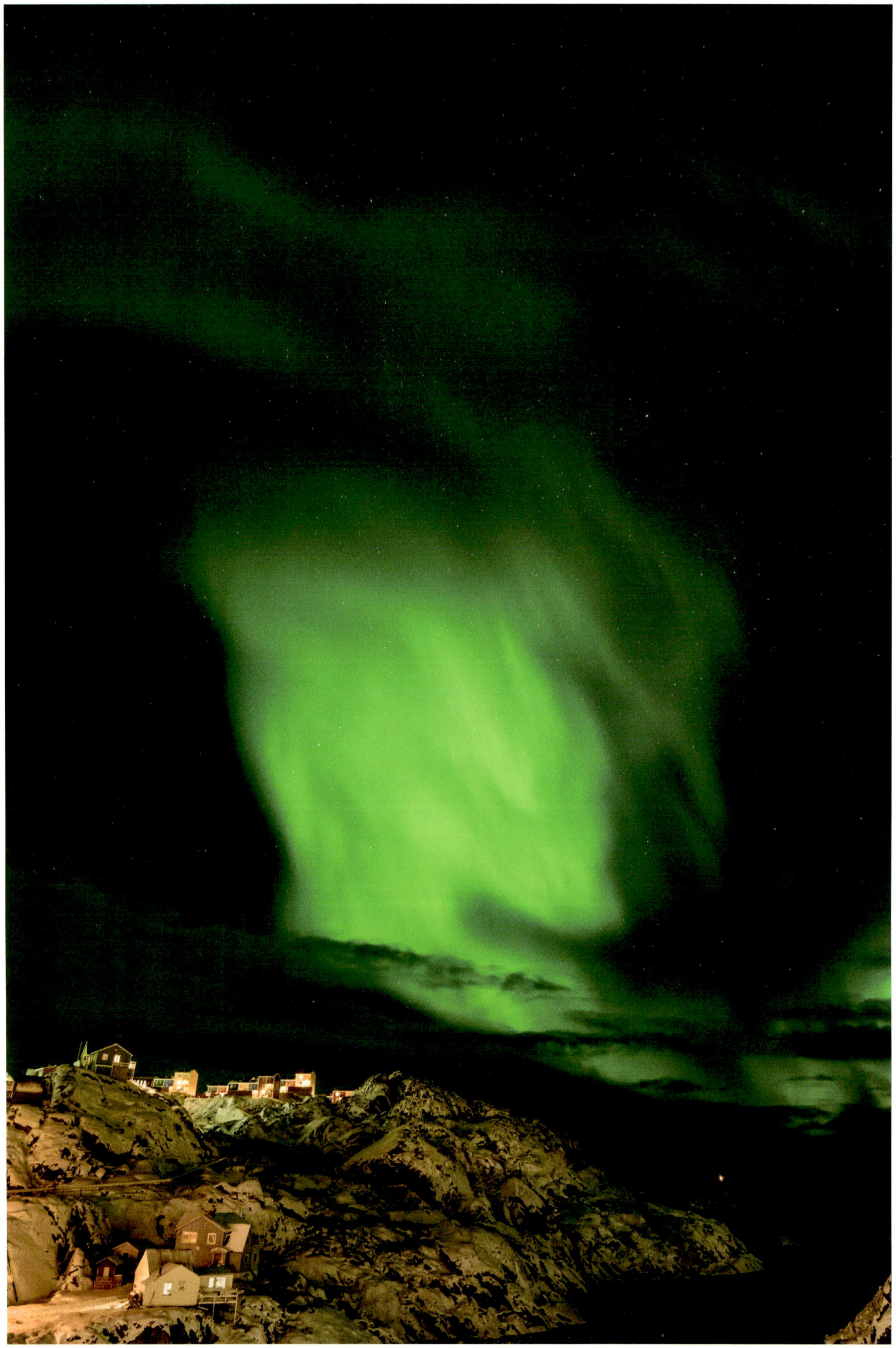

SISIMIUT 2016
Aurora borealis is often seen in Greenland during cold winter evenings. This was a particularly intense night of light movement.

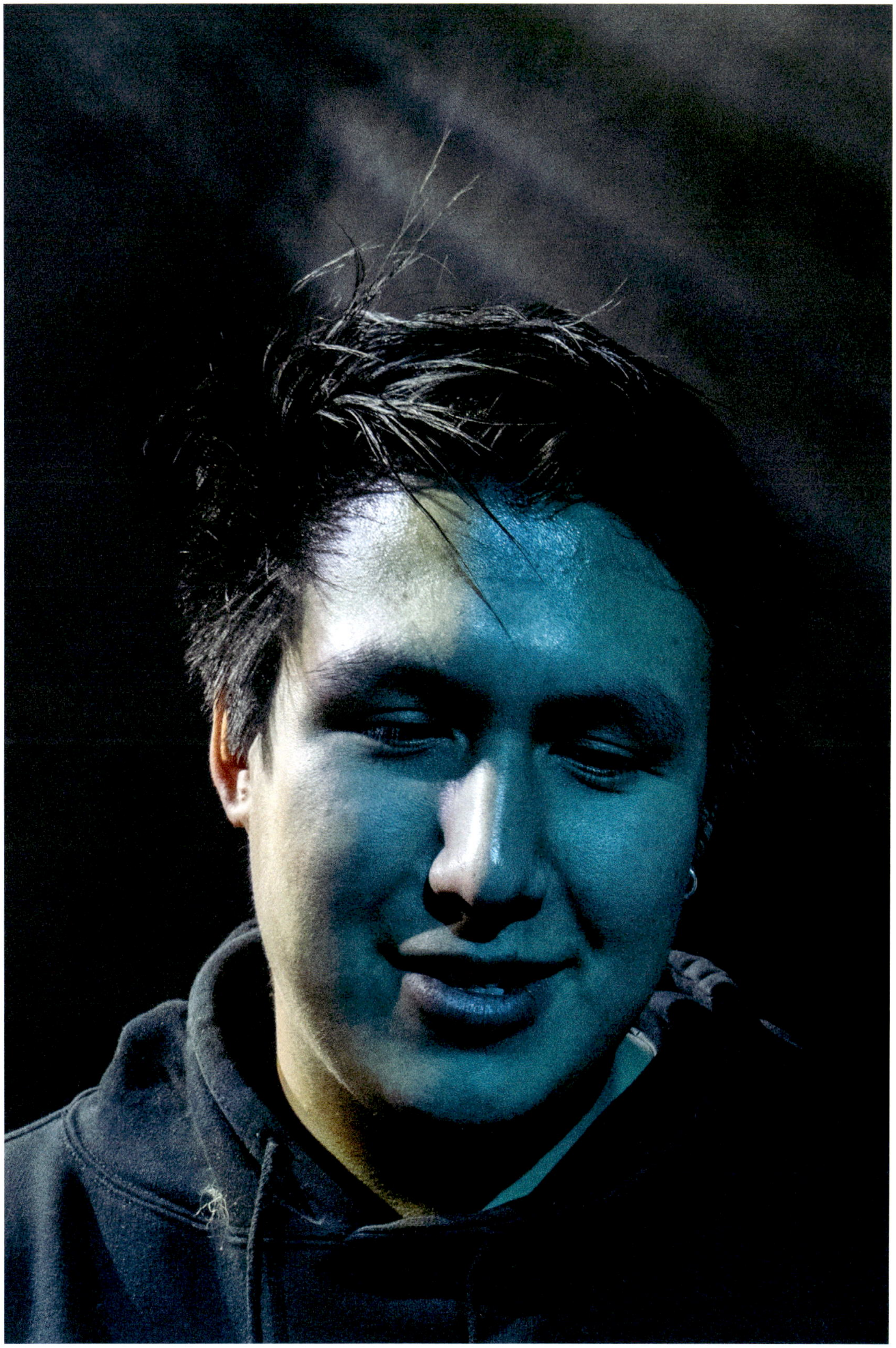

NUUK 2017
Hans-Ole Amossen, music and sound technician for Nalusuunerup Taarnerani production at the Katuaq Cultural Center.

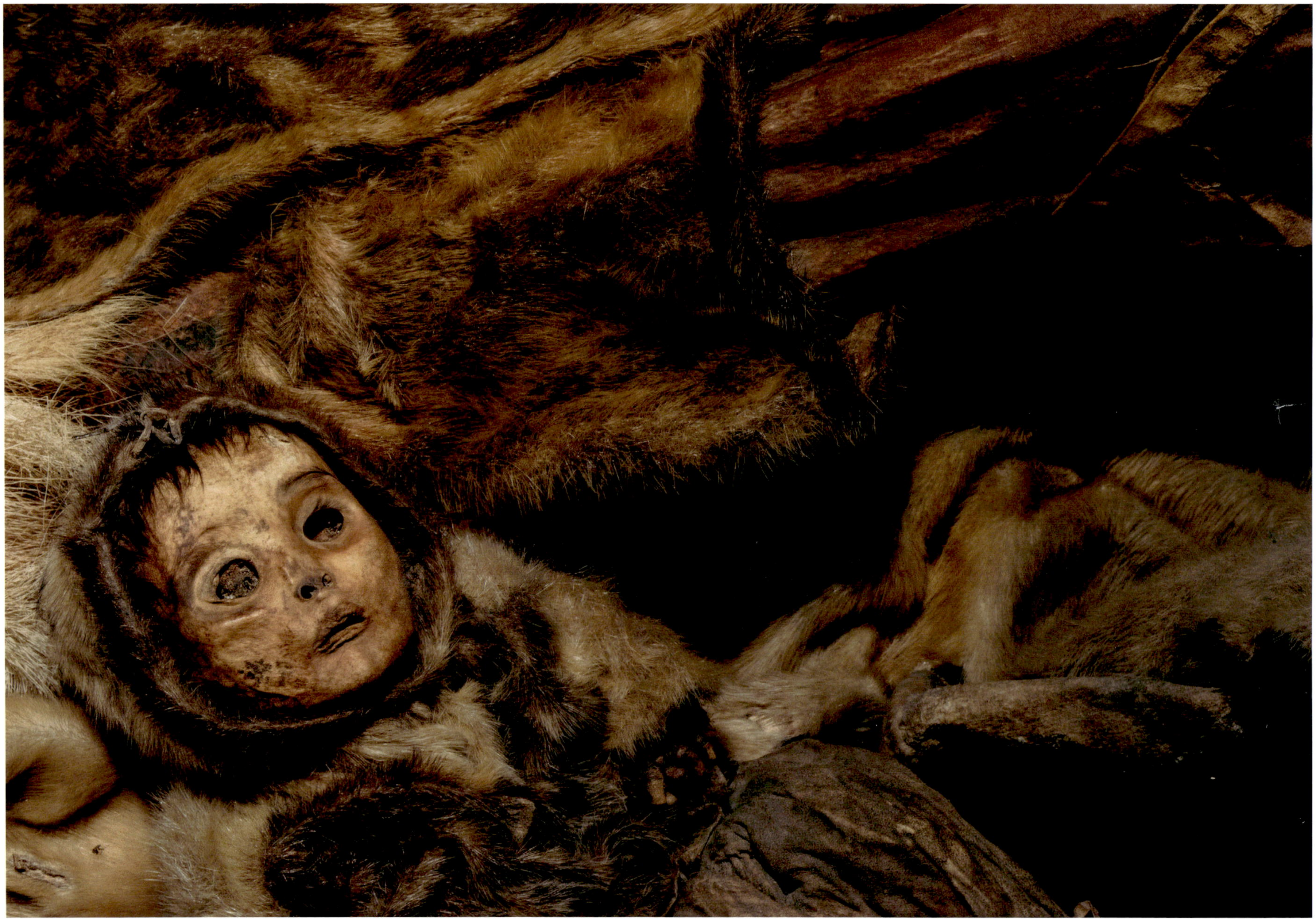

NUUK 2017
Mummified child found at Qilakitsoq, an archaeological site on Nuussuaq Peninsula, on the shore of Uummannaq Fjord.

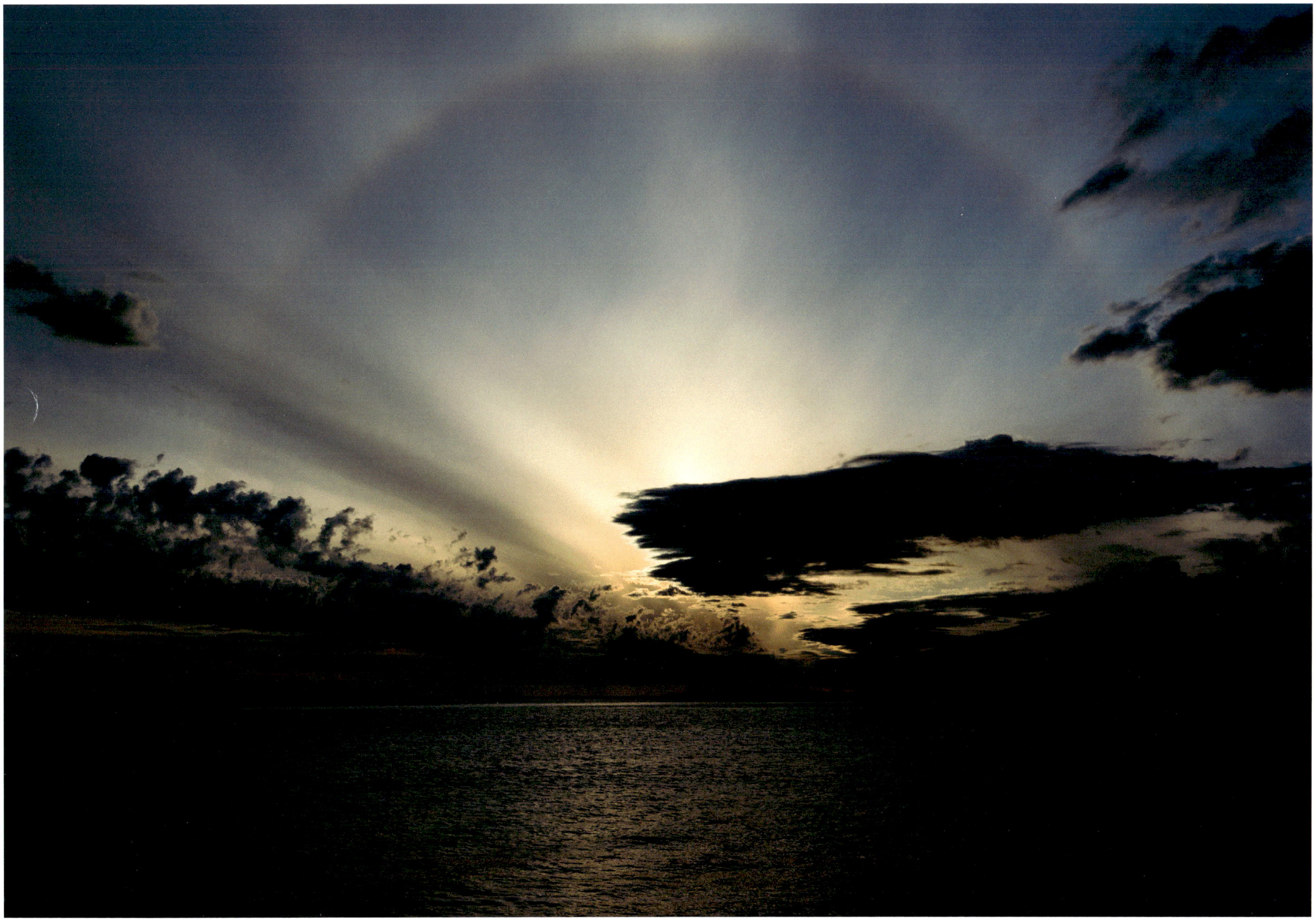

NUUK 2016

Glowing sun dog. Greenlandic myth says when a sun dog appears, it allows the shaman to enter another world to gain spiritual knowledge.

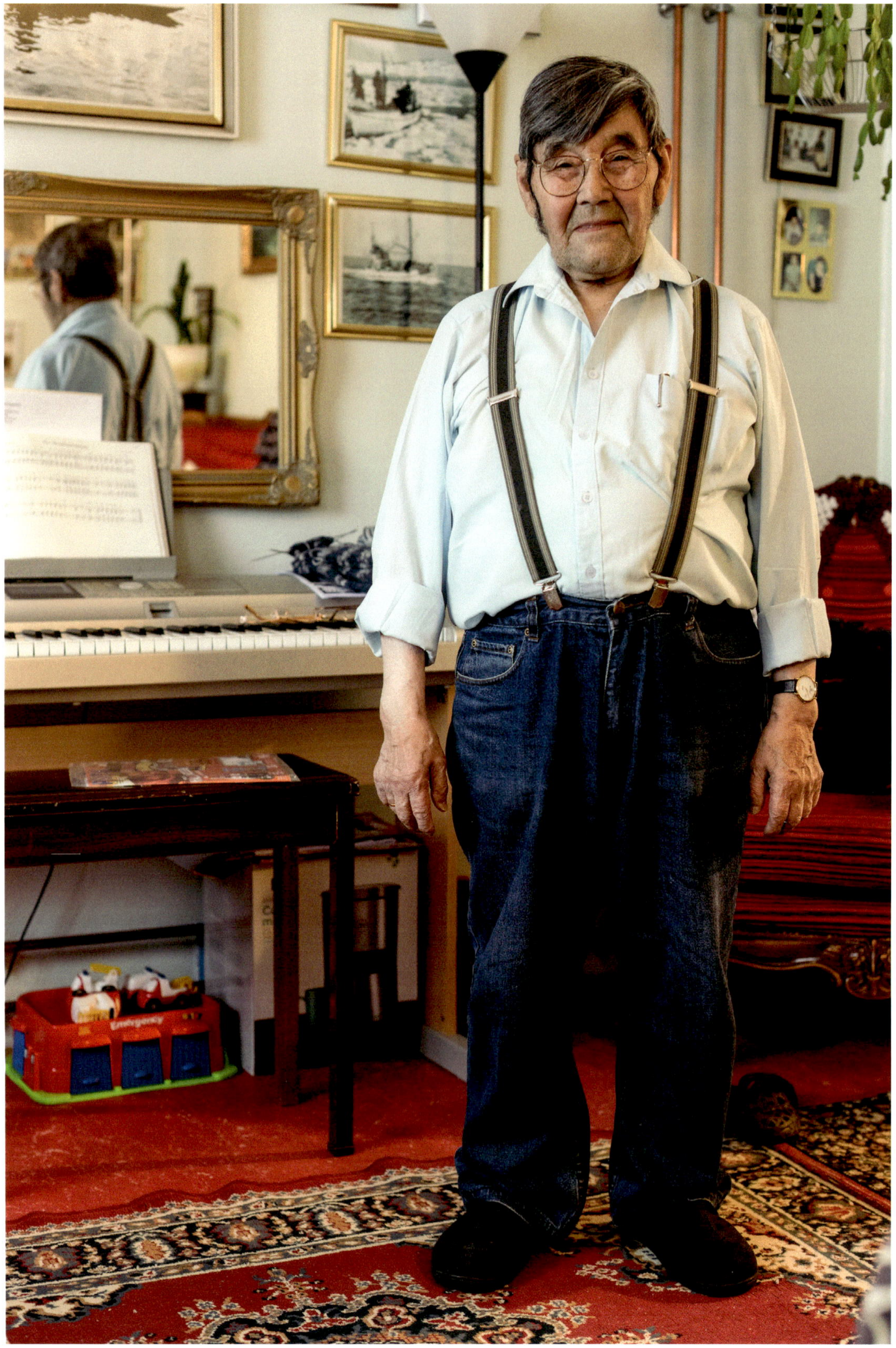

SISIMIUT 2017

Ship captain, sailor, fisherman, and writer Kristian Johansen at home with photos of his history.

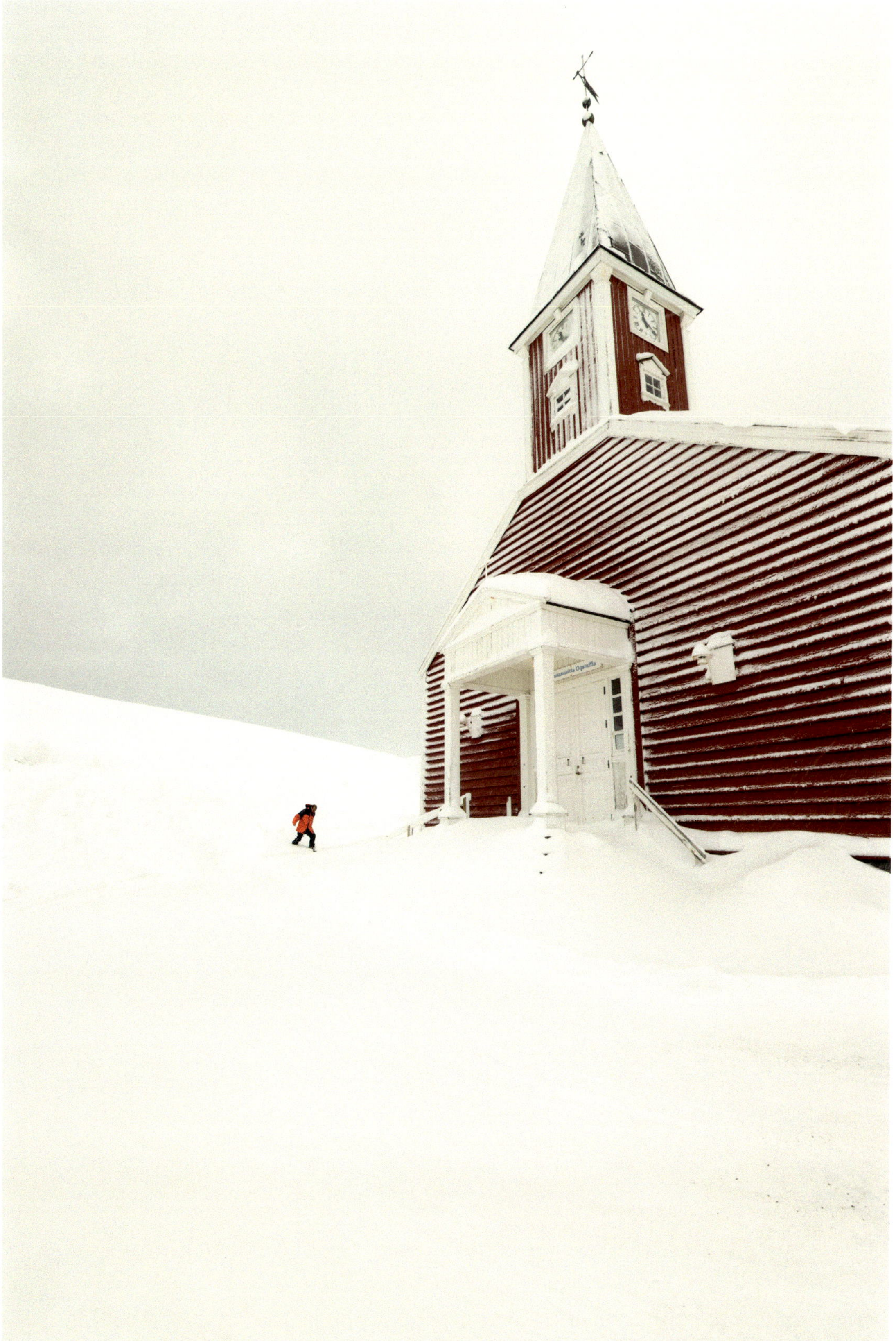

NUUK 2017

Snowstorm envelops the historic Church of Our Saviour, built in 1848–49.

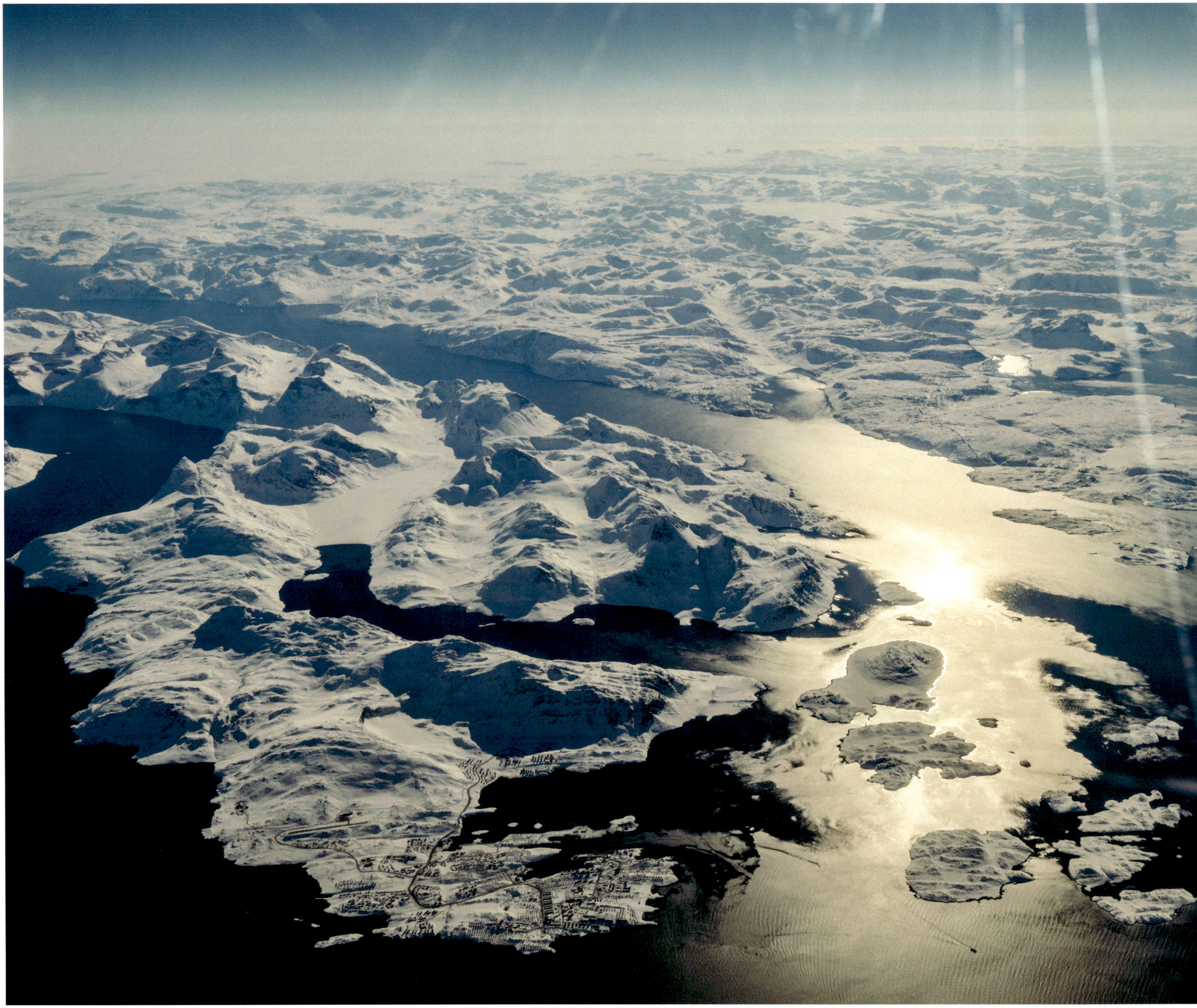

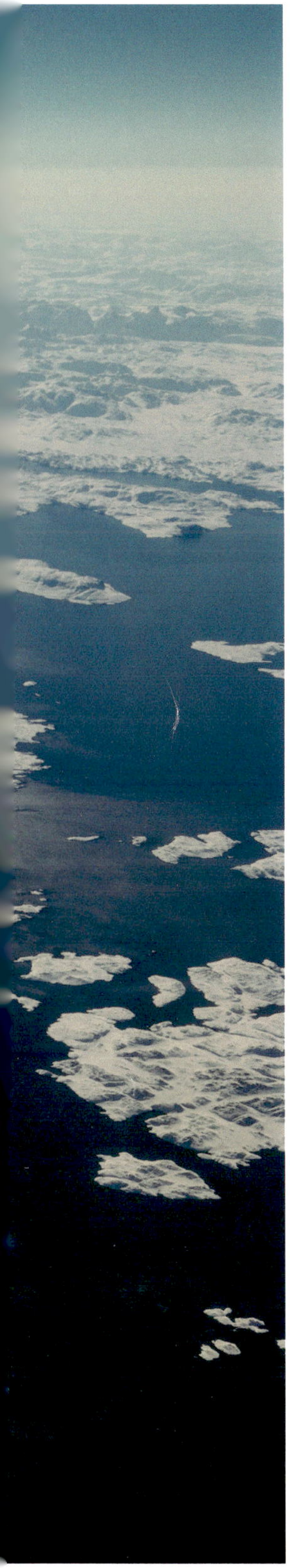

NUUK 2017

Nuuk, Greenland's capital, is surrounded by the extensive Nuup Kangerlua fjord system, the largest such system in western Greenland.

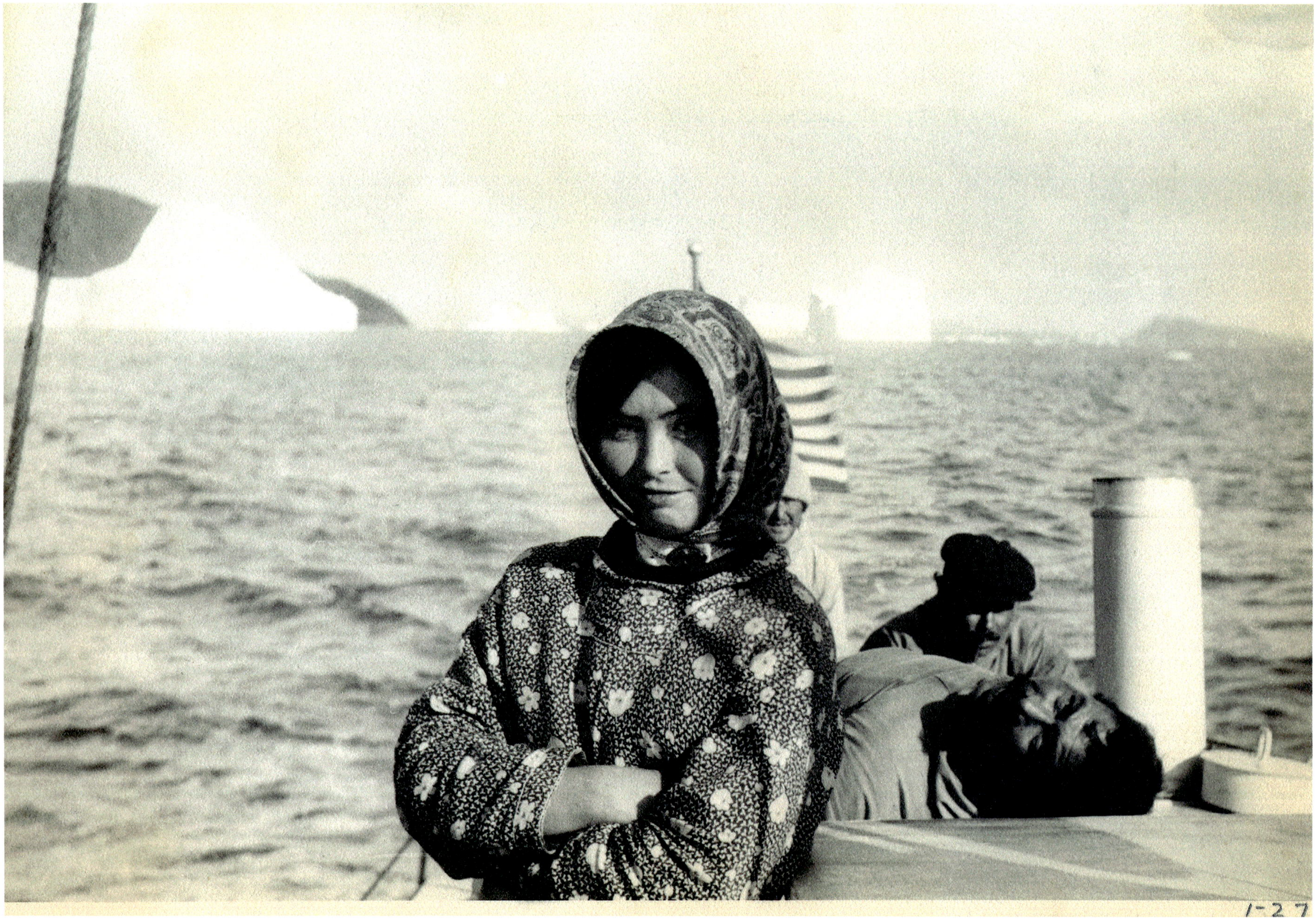

ILLORSUIT SOUND *ca.* 1931
Salamina with Knut Nielsen, Hendrik Qvist and an unidentified passenger sailing with the United States flag in the background; maybe to honor Kent?
*RK-631, Plattsburgh State Art Museum*

# Rockwell Kent's Greenland: Photography, Painting, and Lectures

DENIS DEFIBAUGH

**"A very fine people, the Greenlanders," Rockwell Kent said in a 1934 interview with the *New York Herald Tribune*. "All marvelous physical specimens, the men strong and with great chests; the woman carefree, but a trifle bowlegged. Their personal lives are free and unrestrained, they are very polite and happy, and they live entirely on meat. Their social life is a riot."[1] Kent's affinity for Greenland was palpable.**

Kent had an opinion about everything and seldom hesitated to express his perspective when the opportunity arose. He often was forthcoming with comments and writings about his paintings, describing their success or failure and his approach to creating them. It is odd, therefore, that any self-reflection about his photography—descriptive or critical—is rare. He was an avid photographer who would document a story or adventure for his needed purpose, but never treated photography as an art medium; this made his photographic images quite different from his paintings. His photographs exist mainly as a document of his travel experiences; to provide visual material—lantern slides—for his lecture tours; and as a reference for his prints, drawings, and paintings.

Travel and adventure motifs often are the subjects of Kent's photographs; at times, his approach is to photograph people in their environment for scale, just as he might use them in his paintings. In Illorsuit, he made many photographs, including portraits of people, group photos, documentary photographs of his dog sledging, plein-air painting records, landscapes, seascapes, and social gatherings. All of his Greenland photographs were made with black and white film using a state-of-the-art early model Leica I camera. After editing, the negatives were sent to the Joseph Hawkes Studio in New York City, where they were made into black and white glass lantern slides and hand colored in final preparation for his lecture tours.

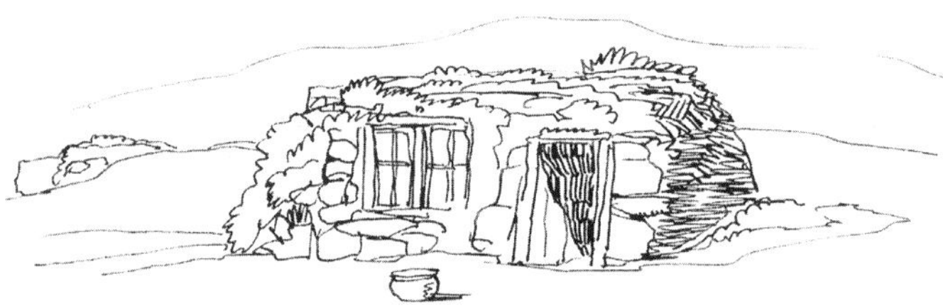

Kent was skilled at producing intimate images, such as *Salamina on the Boat with Henrich Qvist and Knut Nielsen, RK Photo 631.*[2] Salamina commands the photograph with her centered composition and confident, spontaneous smile. The layered figures are "addressing" the camera, possibly as Kent asks for their attention. The photograph is reminiscent of 1930s or '40s Hollywood portraits or fashion photographs of models and movie stars by Steichen or Hoyningen-Huene. Surprisingly, the Stars and Stripes is flying on the mast just behind Salamina in the photo. Why is a Greenlandic boat in 1930s Greenland waters—likely the Uummannaq Fjord—flying the flag of the United States of America? Maybe the crew or Kent himself thought it was appropriate. Kent was fond of flying the flag at his house or during his travels.

Another photograph with equally dramatic composition is *Dogs and Passengers on Board, RK Photo 5866*. Here, Kent takes a tight composition with foreground emphasis to show a dog-filled boat and hints of those aboard without identifying anyone; one passenger in sunglasses lying against a mass of fur peers at Kent though darkened lenses. The photo perfectly represents the tight quarters passengers endured when traveling through the fjords or along coastal Greenland, with dogs, people, supplies, and sledges shuttled together between ports. In the photo, snow is packed along the hull, and dogs and passengers alike are bundled to keep warm. One can almost feel the cold in this unusual aesthetic and composition for Kent.

Viewers of his photographs soon realize that Kent uses people to give a sense of scale to the enormous mountains, icebergs, and broad landscapes. In addition, Kent's depictions of Greenlandic people typically show them going about their business rather than posing, a style markedly different from Kent's other photographs, which are more intimate and posed to show the character of the subject or to document the actions or events that occur as he photographs.[3] Kent's paintings and graphic art represent both Romanticism and the influence of William Blake; both expressed divine and celestial spheres in their works. Kent used the representation of man as a small grain in the universe compared to powerful nature. Like Gauguin, who went to Tahiti in search of "the earthly paradise," Kent immersed himself in a remote culture quite different from New York's, seeking the harmonious co-existence between man and nature.[4]

Kent's photographs provide ethnographic details of Inuit culture and the North. Clothing, faces, homes, the hunter's workplace, dog sledging, and portraits depicting relaxation add to a story of robust life in Illorsuit. The hand coloring of the lantern slides, with spots of brown dye covering the faces of many Inuit subjects or the overly expressive color of some clothing, adds hints of Romanticism to the images. When the hand-coloring dyes are added conservatively and accurately, they provide credible and believable results.

Kent shared his photographs at speaking engagements that he undertook to supplement his income and to fund future travel projects. The Colston Leigh Bureau of Lectures and Entertainment scheduled one such tour for Kent—a rigorous multi-city, 82-day trip—to raise funds for an upcoming 1937 trip to Greenland. Kent's speaking fee for a San Diego lecture that year was $250, for instance. Kent's lantern slides were quite popular at these lectures and perfectly complemented his amazing energy and engaging platform presence.

"The Rockwell Kent lecture was splendid [and] his pictures were delightful,"[5] the Women's Club of Evanston, Illinois, reported. At the University of California, Berkeley, Kent "gave his usual delightful lecture," drawing upon his strong skills as an orator. "It is unusual," the Berkeley review noted, "to have a lecturer with such a keen sense of humor, and yet not lose sight of the serious side of his subject, which he so soundly presented."[6] Likewise, the Evanston Club noted with apparent satisfaction that, "Our clubhouse was full and we turned away a number of people."[7]

Though this lecture tour was immensely successful, and Kent's affection for Greenland and its people genuine, his 1937 return trip to that country unfortunately never happened.

**Notes**

1 "Rockwell Kent off Wednesday for Greenland," *New York Herald Tribune*, July 6, 1934.

2 The "RK" designation identifies a photograph as part of the Rockwell Kent Collection at the Plattsburgh State Art Museum of the State University of New York at Plattsburgh.

3 "Arts and Activities," June Summer 2001, 33.

4 Alexandra A. Sitnikova, "The Concept of the 'North' in the works of Rockwell Kent," *Journal of Siberian Federal University, Humanities and Social Sciences 8* (2014-7), 1363.

5 Brochure with commentary, Scott Ferris Collection.

6 Ibid.

7 Ibid.

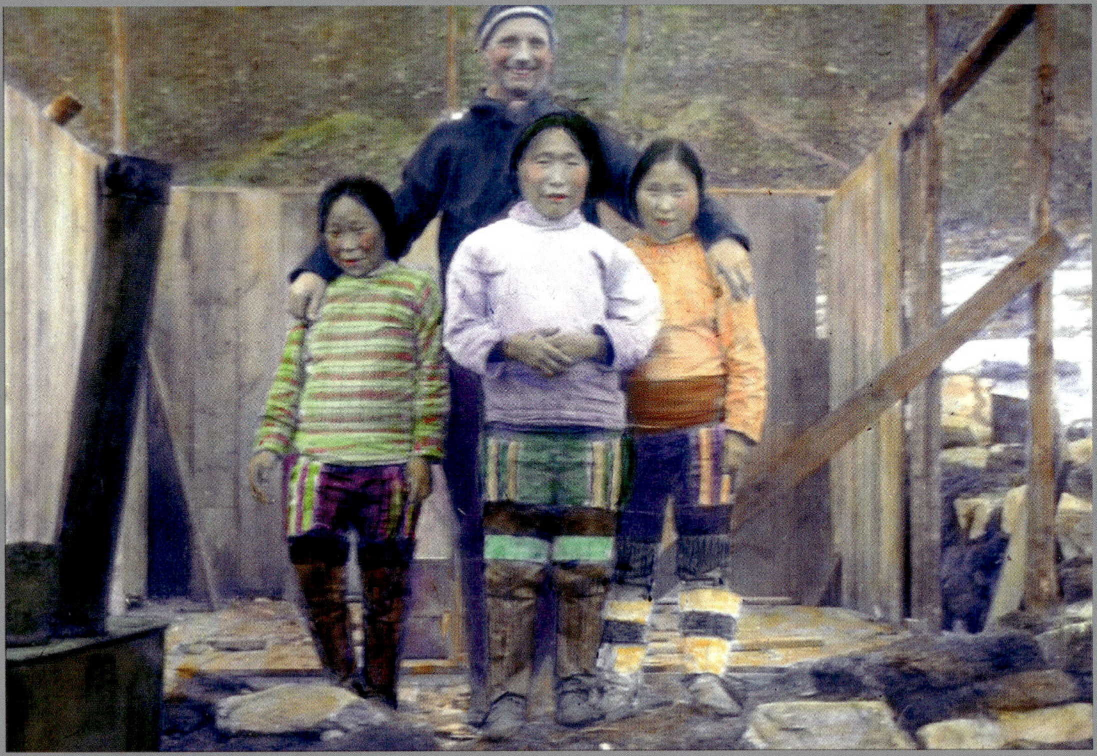

*ca.* 1931
A beaming Kent towers over his Inuit women friends. *RK-5865, Plattsburgh State Art Museum*

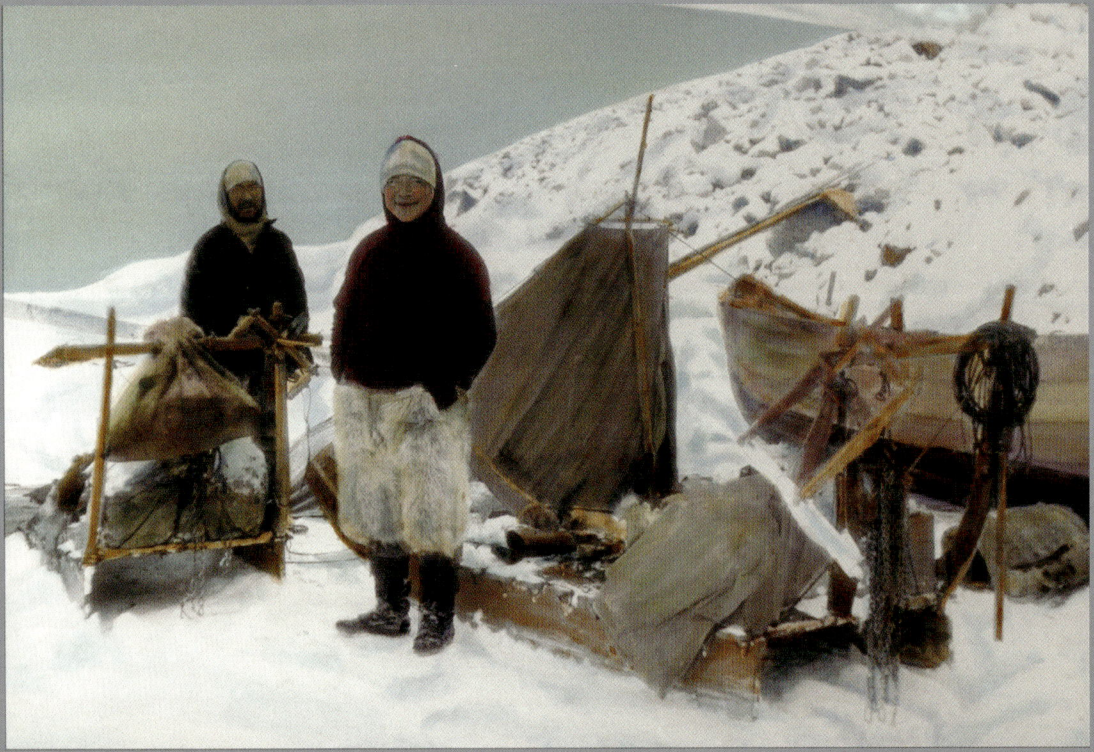

*ca.* 1932
Martin Zeeb (right) with David Lövstrom, who was hired by Kent to take care of his dogs and to hunt for the Kent household. *RK-5838, Plattsburgh State Art Museum*

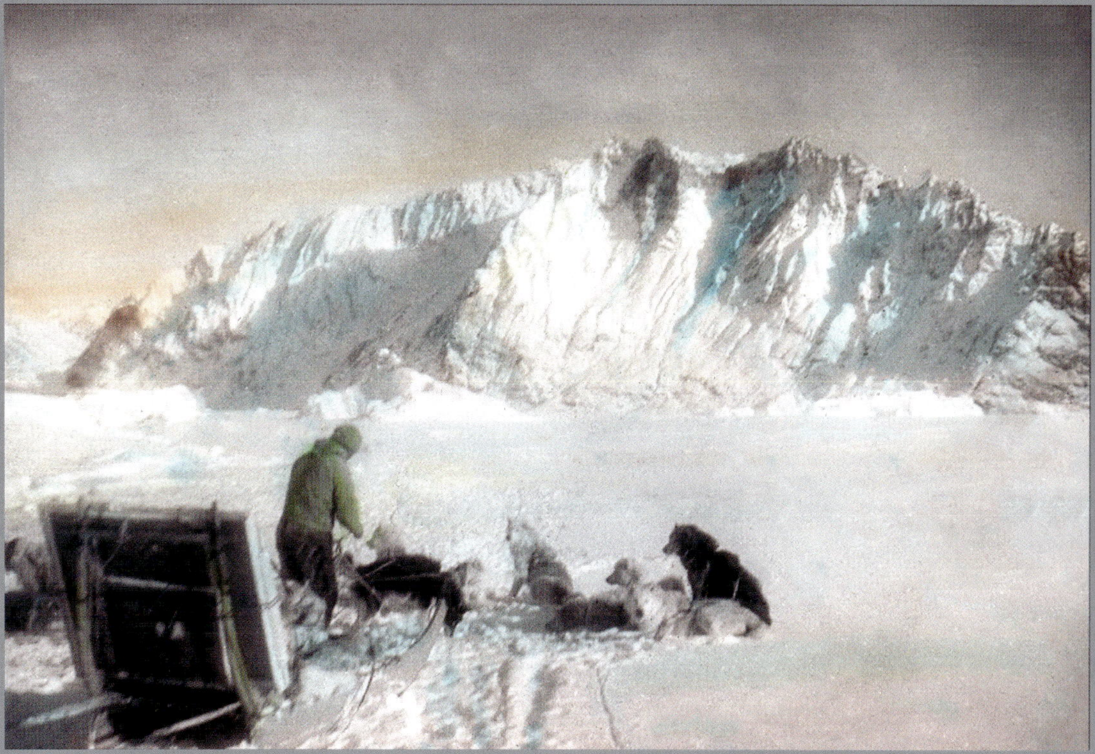

*ca.* 1932
Kent preparing to paint a winter scene on the frozen Sound. Illorsuit. *RK-5868, Plattsburgh State Art Museum*

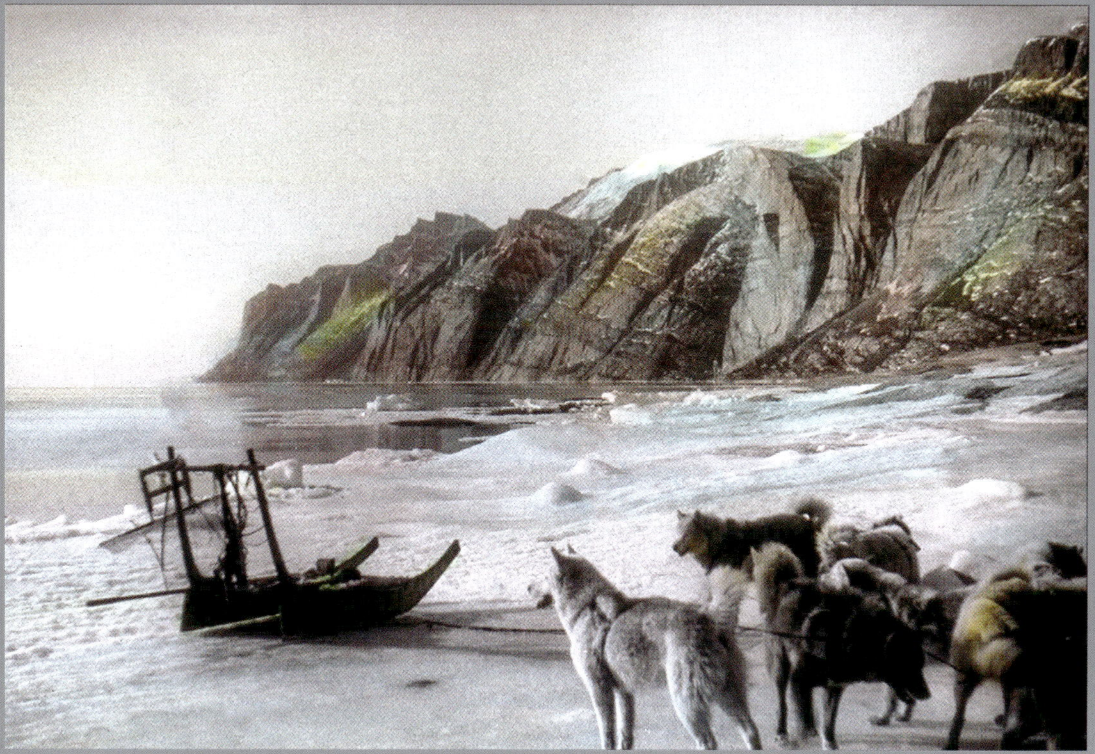

*ca.* 1932
Dogs and sledge at the edge of the water. *RK-5870, Plattsburgh State Art Museum*

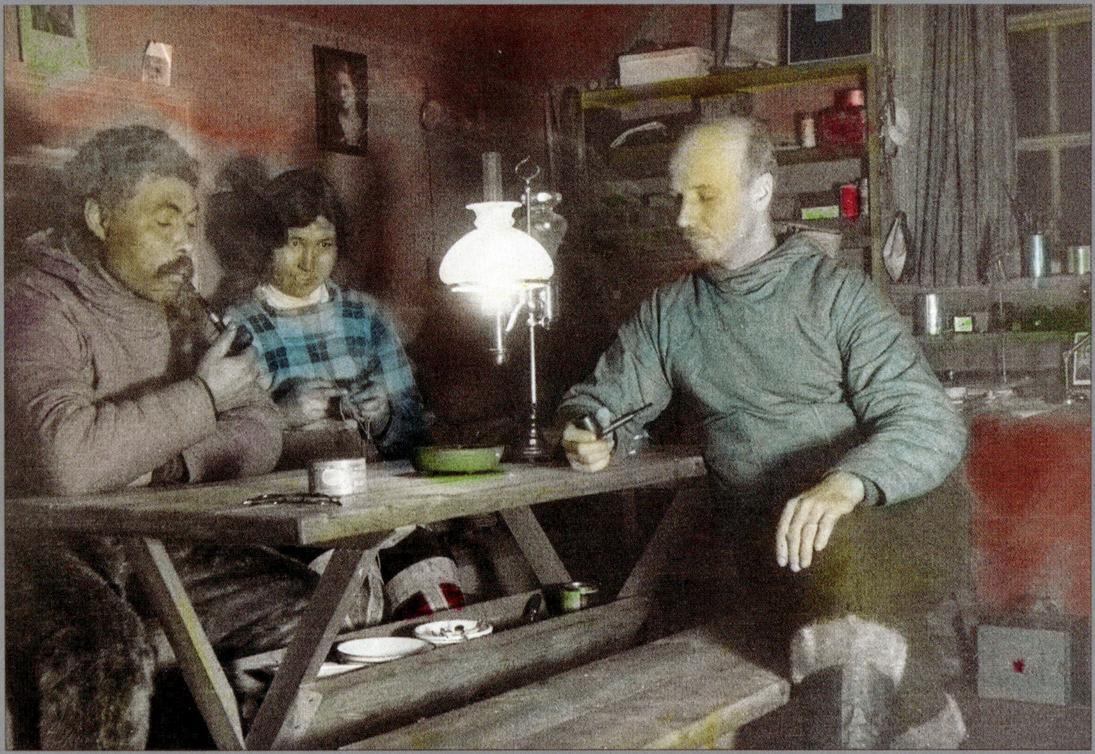

*ca.* 1931
Knut Nielsen, Salamina, and Kent in his cozy one-room home. *RK-5857, Plattsburgh State Art Museum*

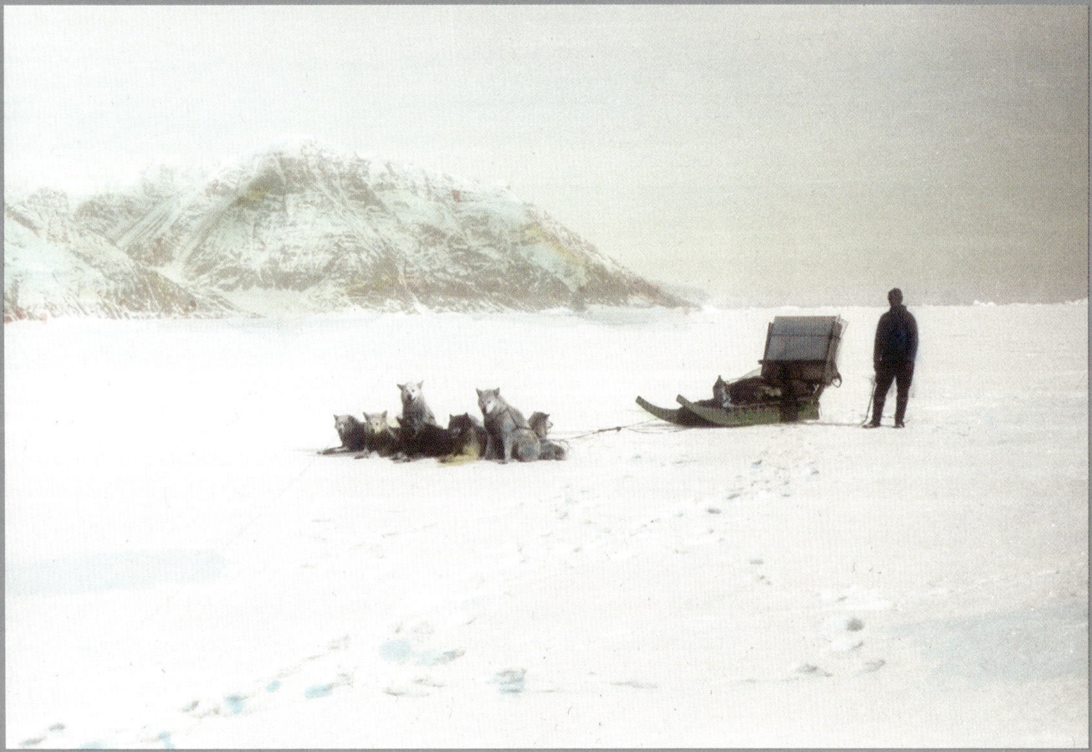

*ca.* 1932
With paintings strapped to the sledge, dogs and Kent are ready to return to Illorsuit. *RK-5839, Plattsburgh State Art Museum*

*ca.* 1931

View of the fine home that Kent built on a hillside overlooking Illorsuit. *RK-5832, Plattsburgh State Art Museum*

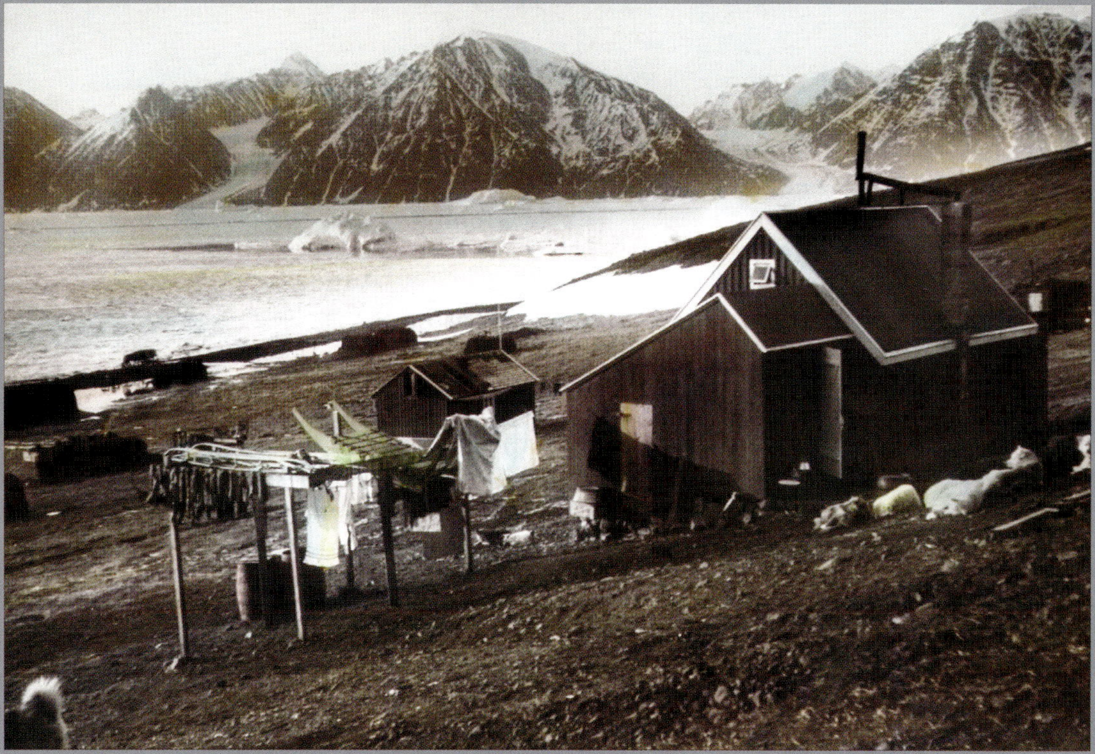

SAATTUT *ca.* 1932

The view of Saattut, Uummannaq Fjord, and the chief's profile on the left of the distant mountains. RK-655, *Plattsburgh State Art Museum*

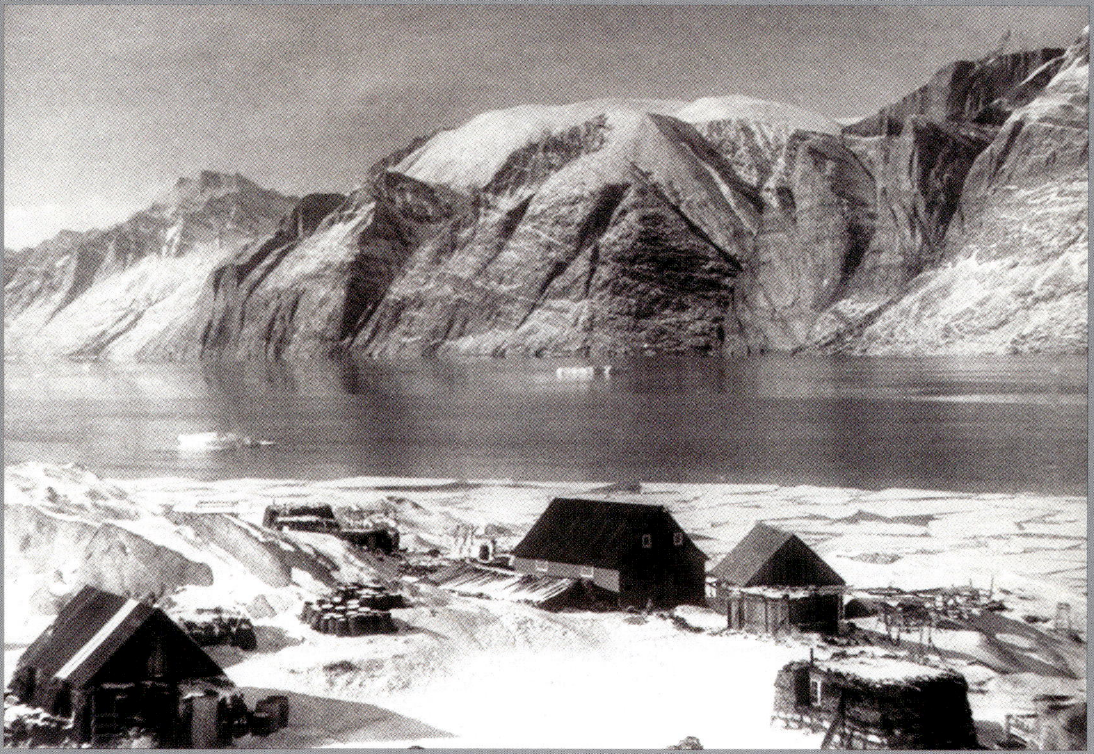

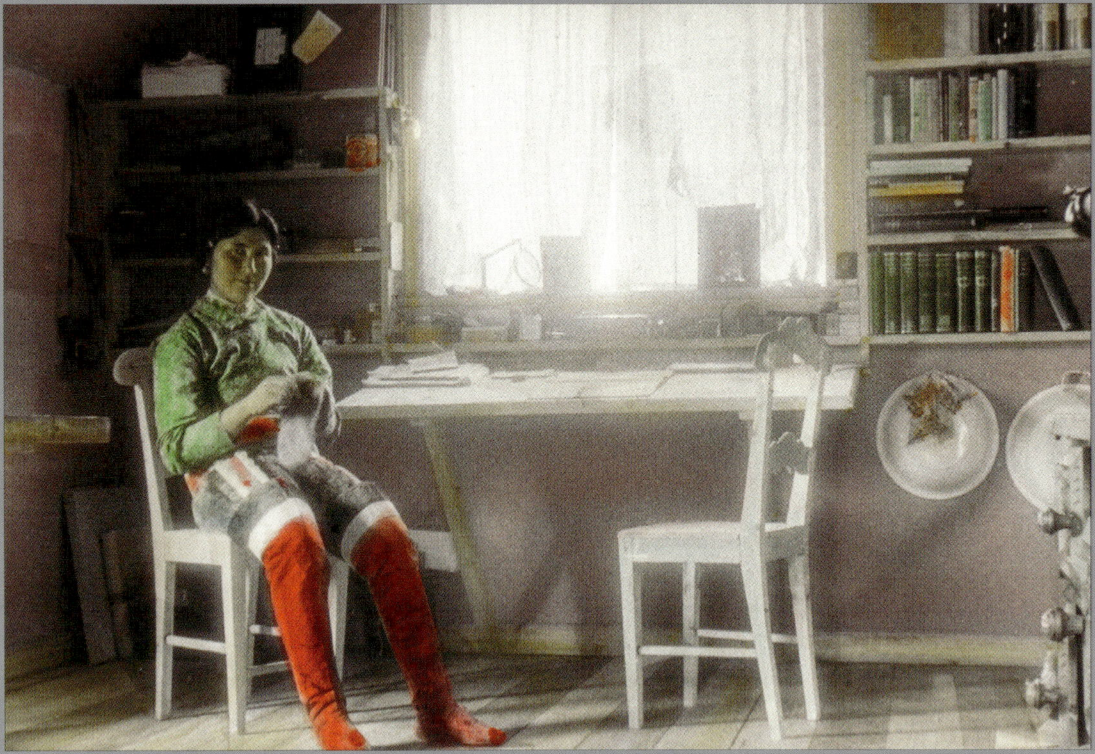

ILLORSUIT *ca.* 1931
Salamina in Kent's home. *RK-5858, Plattsburgh State Art Museum*

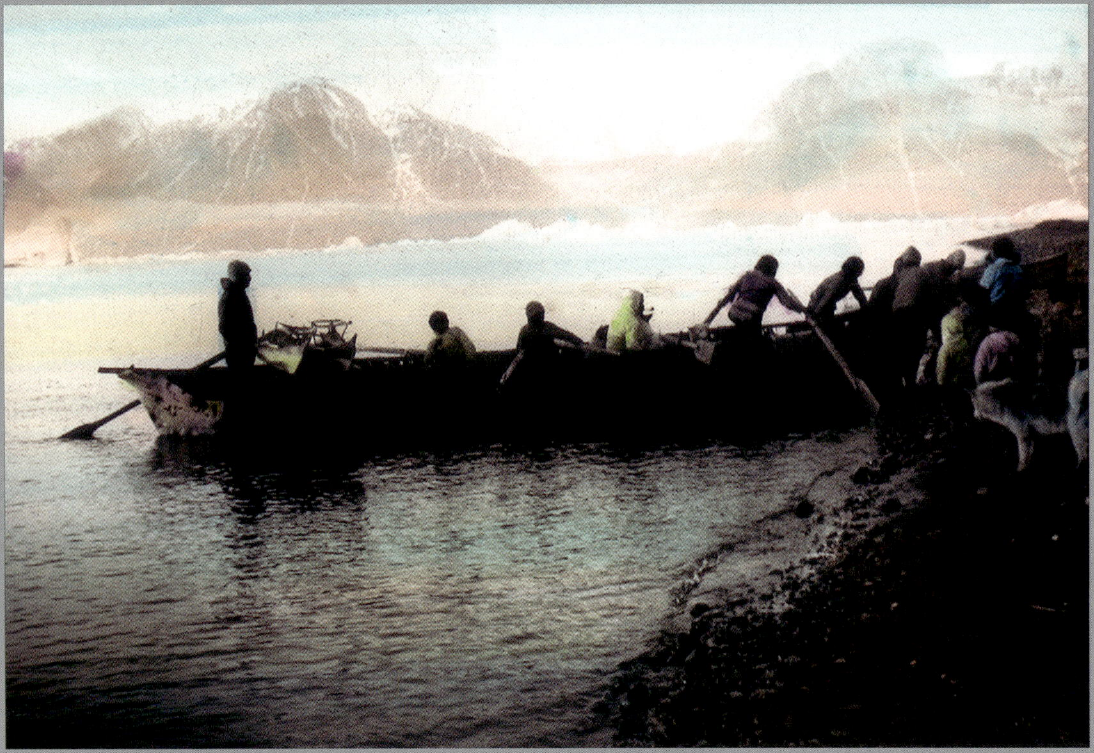

*ca.* 1931
The *umiak* is a traditional Inuit seal skin-covered boat, also known as "the woman's boat."
*RK-5859, Plattsburgh State Art Museum*

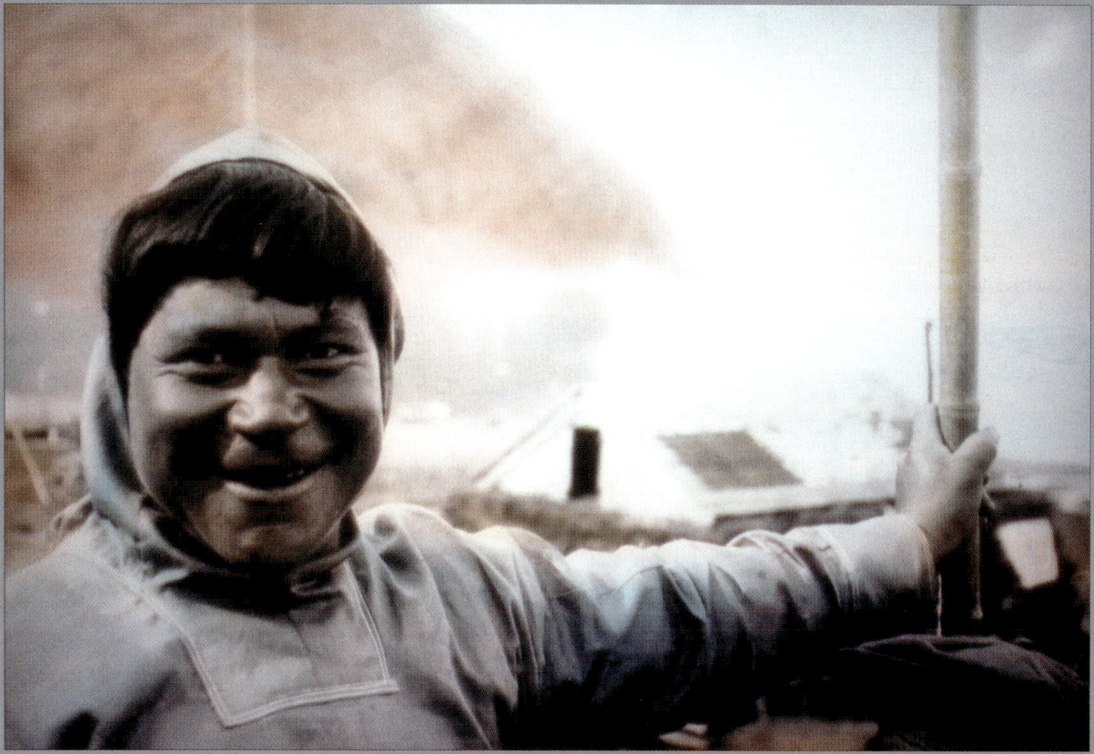

**ILLORSUIT** *ca.* **1931**
Of his friend Martin Zeeb, Kent noted, "He has a face like a full moon, a smile like the rising sun, and tear ducts like Niagara Falls." *RK-5824, Plattsburgh State Art Museum*

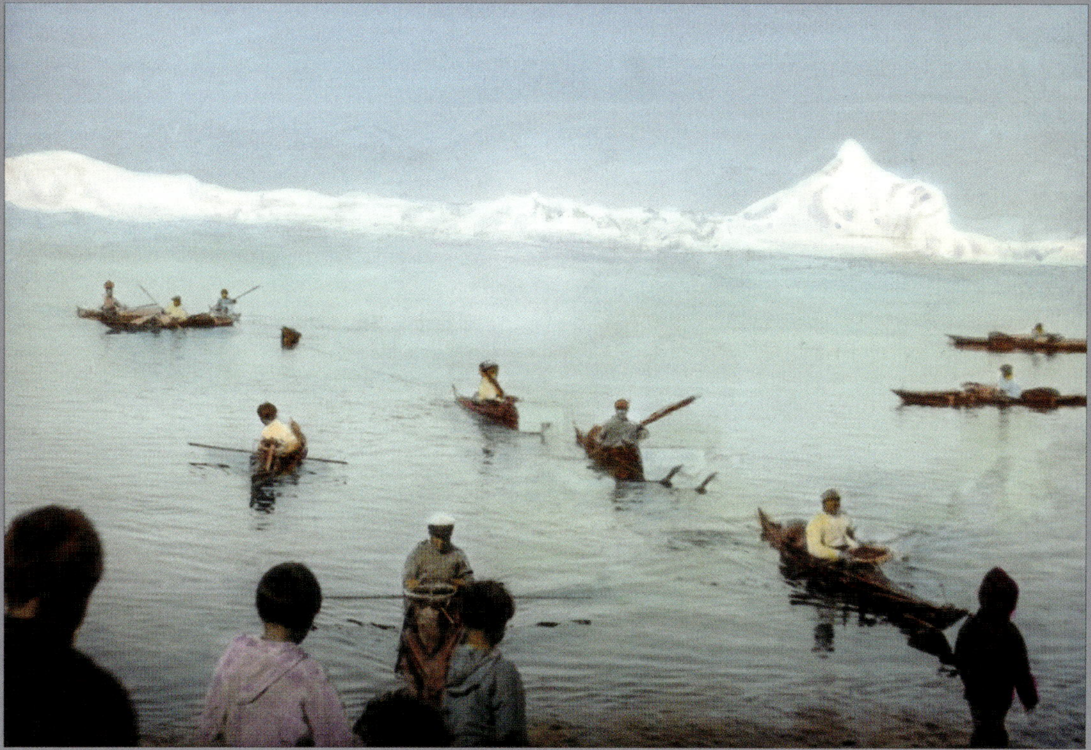

*ca.* **1931**
Kayakers gathering. *RK-5863, Plattsburgh State Art Museum*

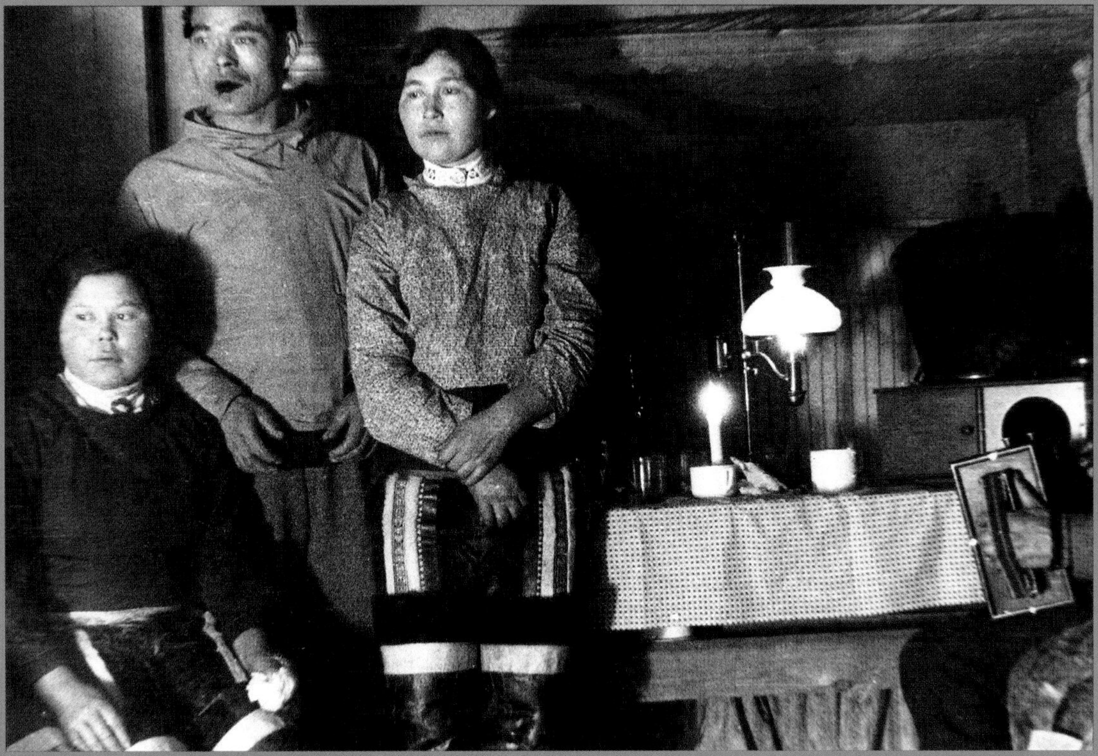

*ca.* 1931
Husband and wife Rudolf Qvist and Marghreta Qvist with an unidentified young friend inside Kent's Illorsuit home. *RK-663, Plattsburgh State Art Museum*

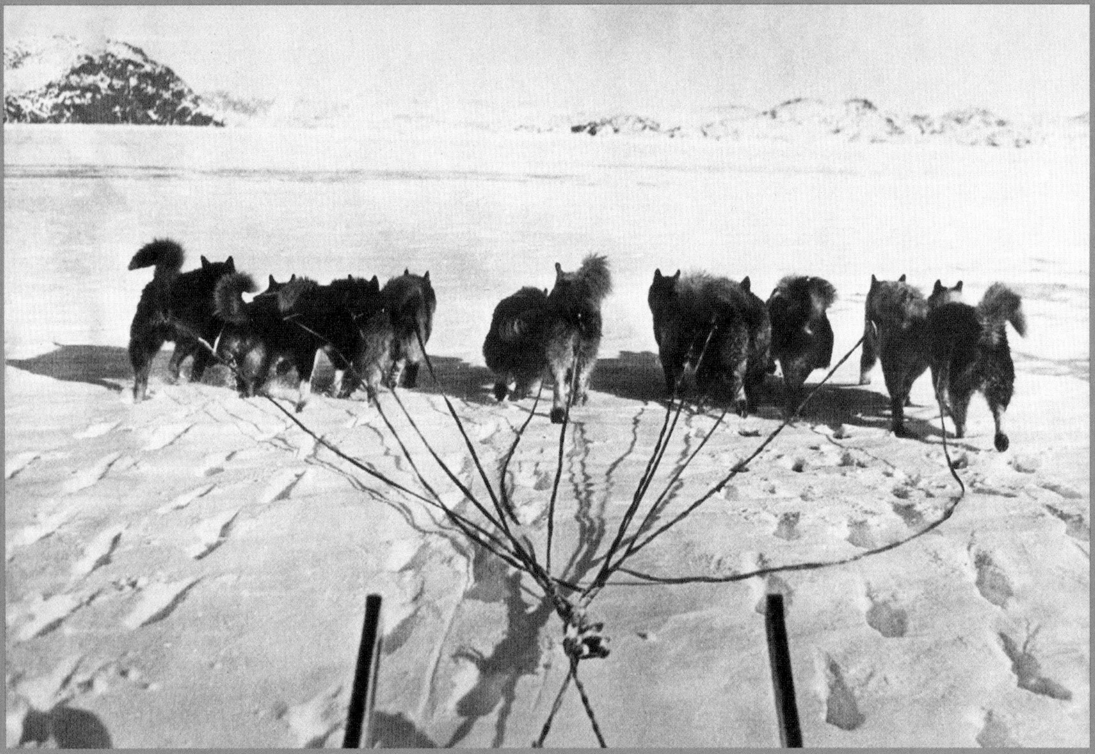

*ca.* 1932
Dogs pulling Kent's sledge across a frozen fjord. *RK-860, Plattsburgh State Art Museum*

**ILLORSUIT** *ca.* **1932**
Group portrait in center of Illorsuit. Most of these buildings are still standing and in use. Amalie Fencher in center dominates the group. *RK-902, Plattsburgh State Art Museum*

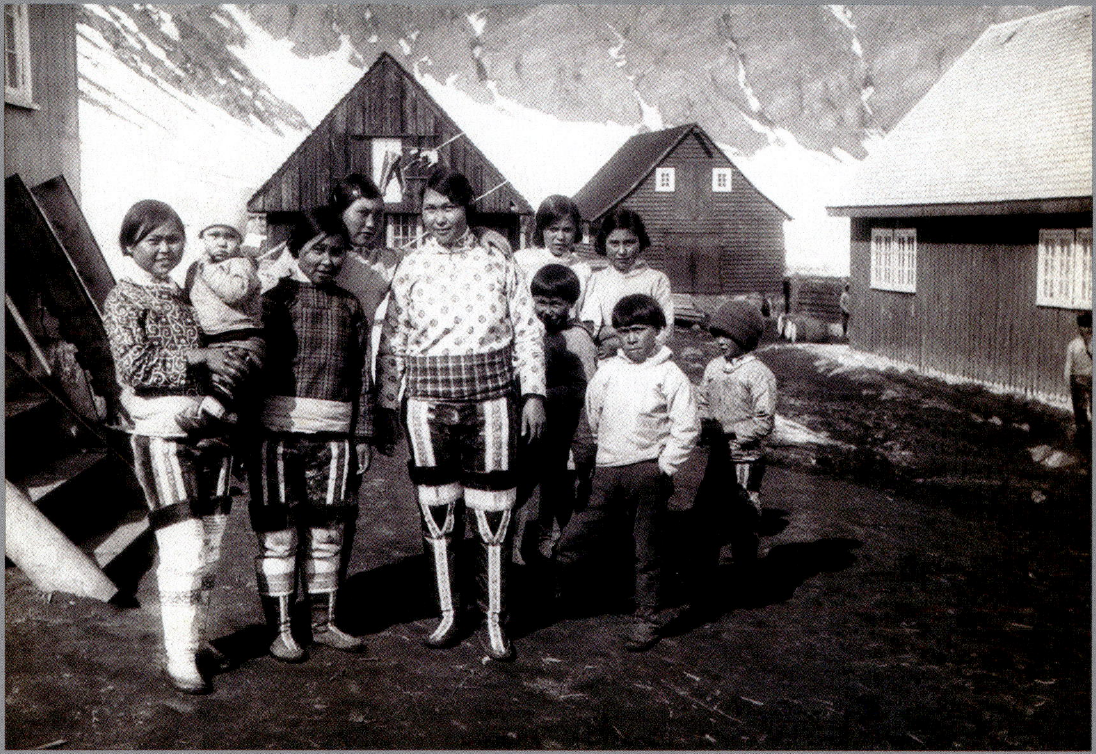

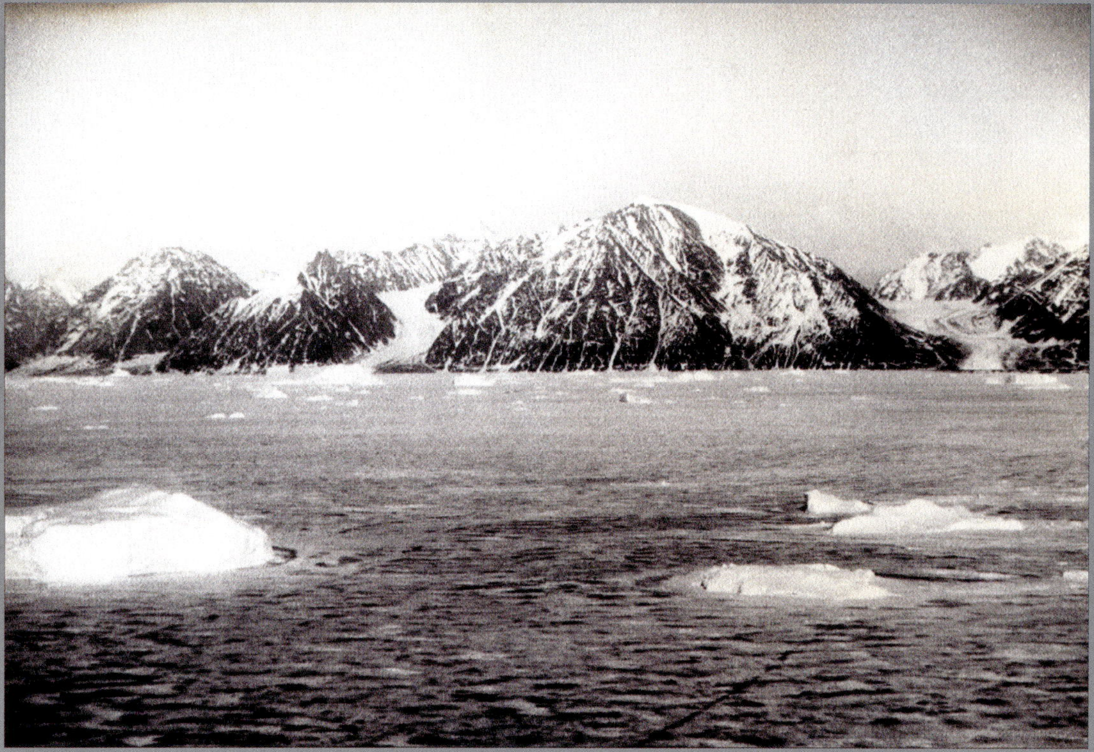

*ca.* **1932**
Illorsuit Sound near the end of winter.
*RK-698, Plattsburgh State Art Museum*

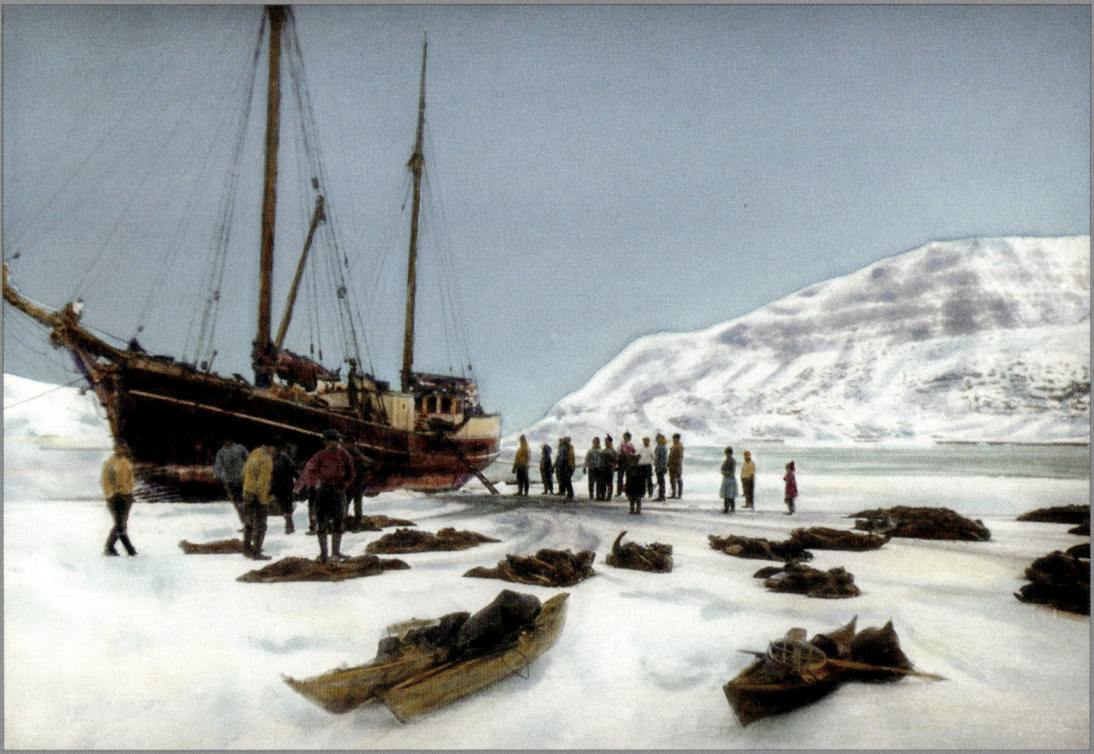

*ca.* 1932
Ship in winter flanked by Inuits and kayaks.
*RK-5861, Plattsburgh State Art Museum*

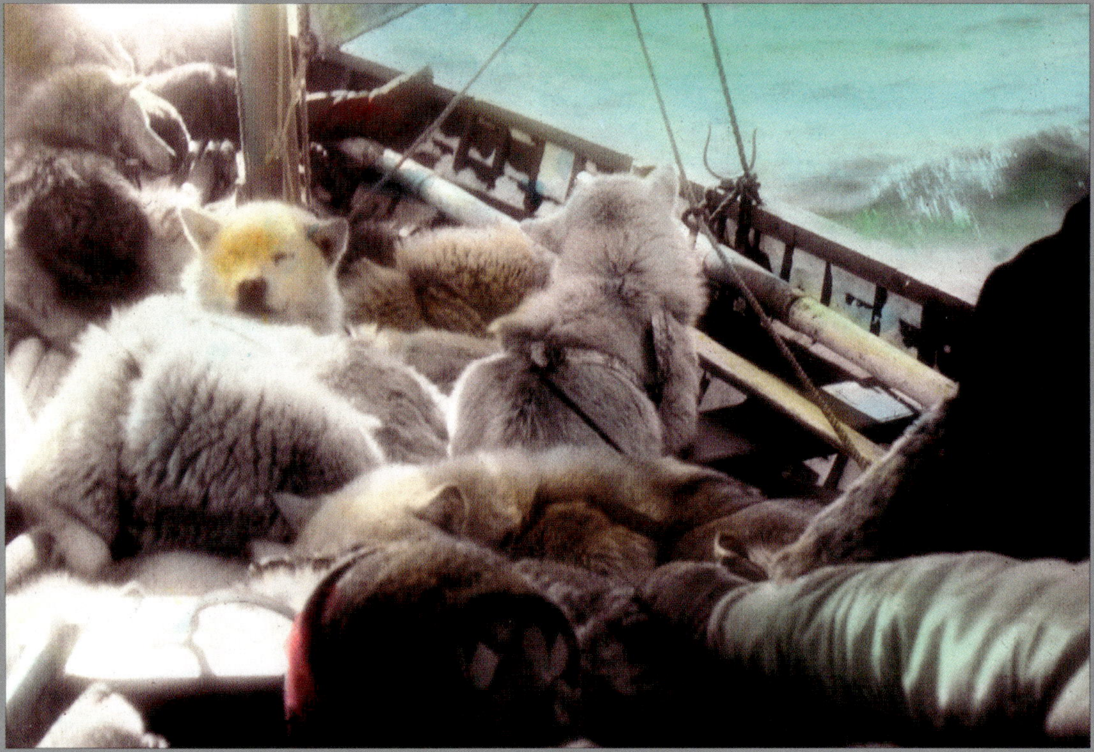

*ca.* 1932
Dogs, sledges and people being transported across a fjord. *RK-5866, Plattsburgh State Art Museum*

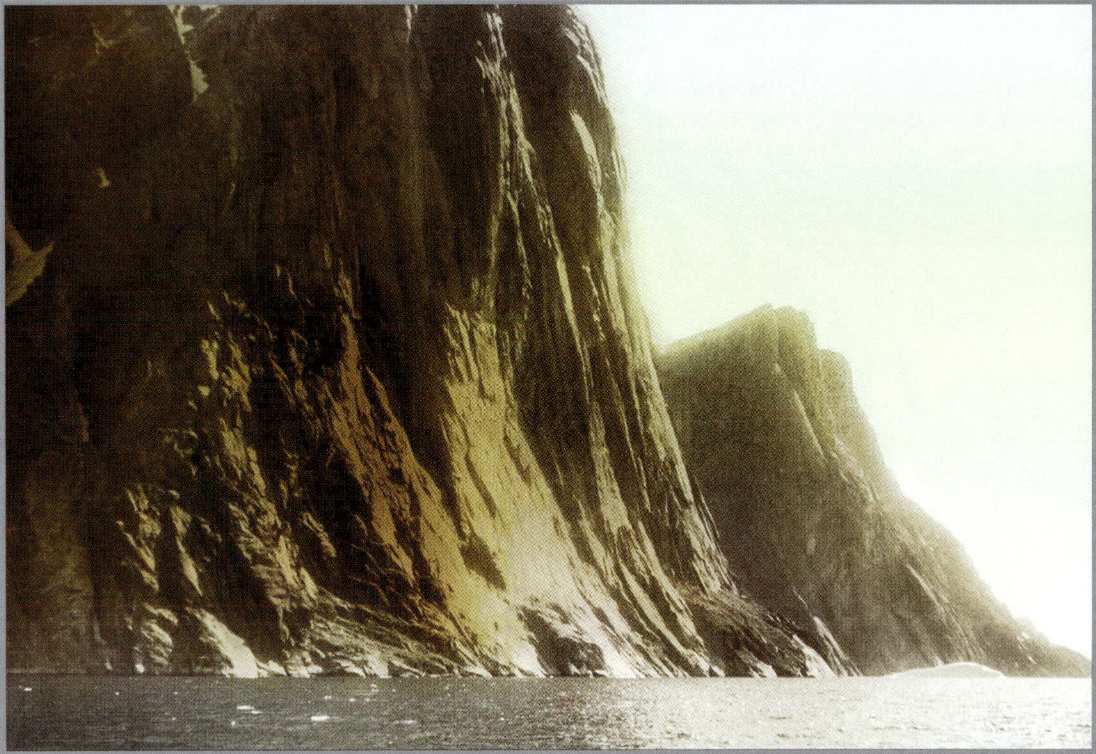

*ca.* 1931
Dramatic cliffs. *RK-5835, Plattsburgh State Art Museum*

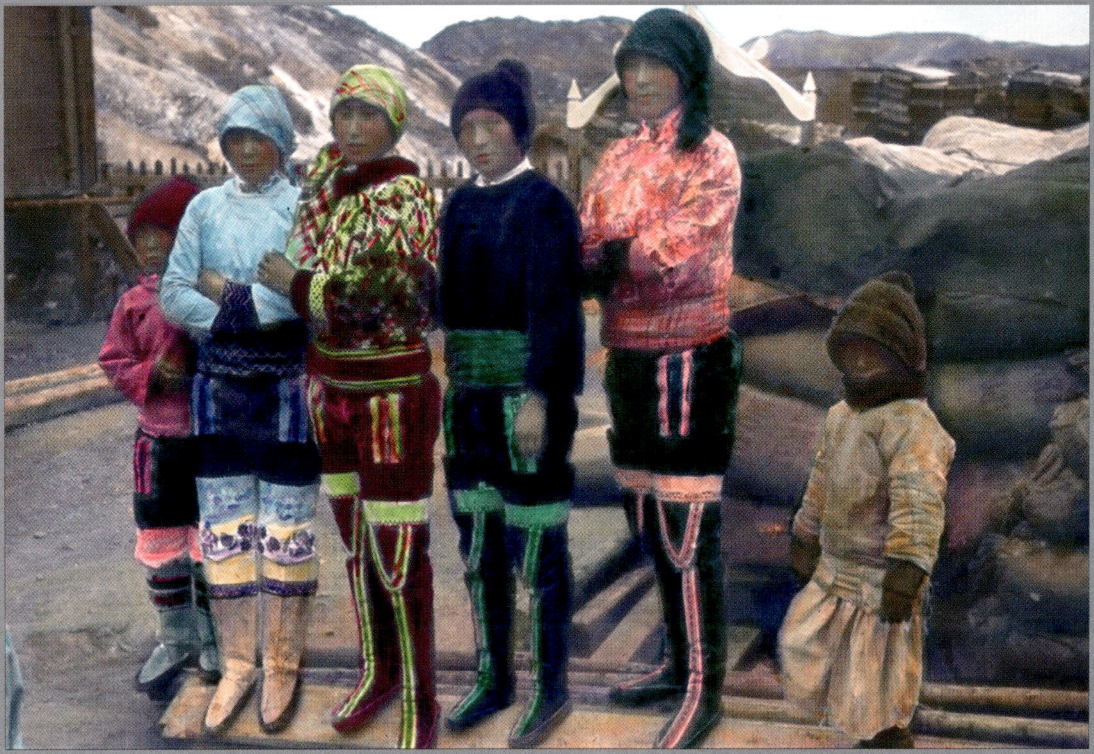

*ca.* 1932
A photograph of exotic Inuit women, each in exaggerated colorful traditional clothing with unusual headgear. *RK-5874, Plattsburgh State Art Museum*

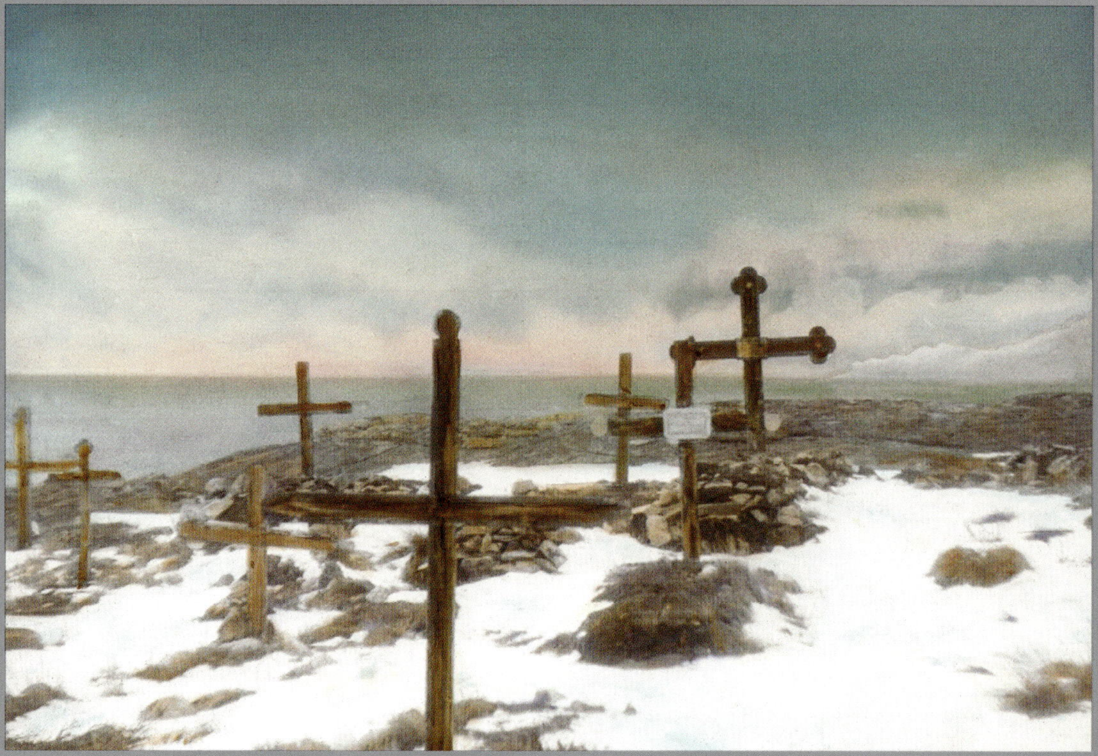

*ca.* 1932
Crosses in what may be the original Illorsuit cemetery. *RK-5828, Plattsburgh State Art Museum*

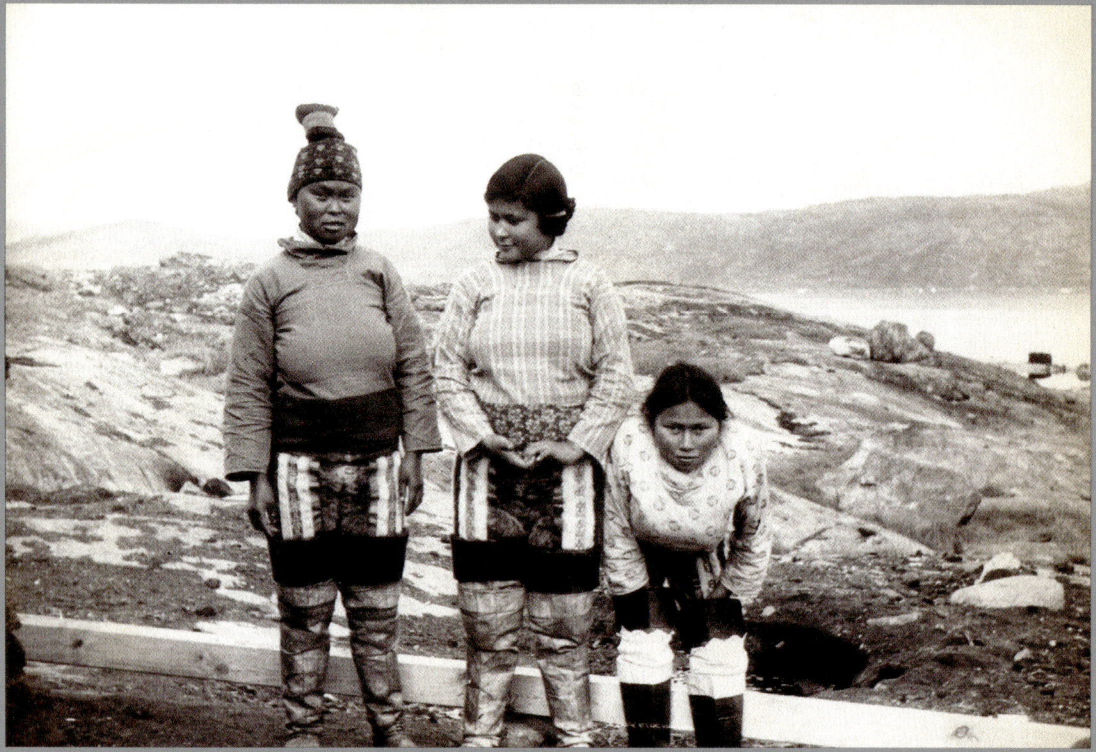

*ca.* 1932
Portrait of Inuit women. *RK-637, Plattsburgh State Art Museum*

*ca.* 1932

Group photo made during Frances Kent's birthday celebration in Illorsuit. The floor of the new dance hall that Kent was building can be seen. *RK-5877, Plattsburgh State Art* Museum

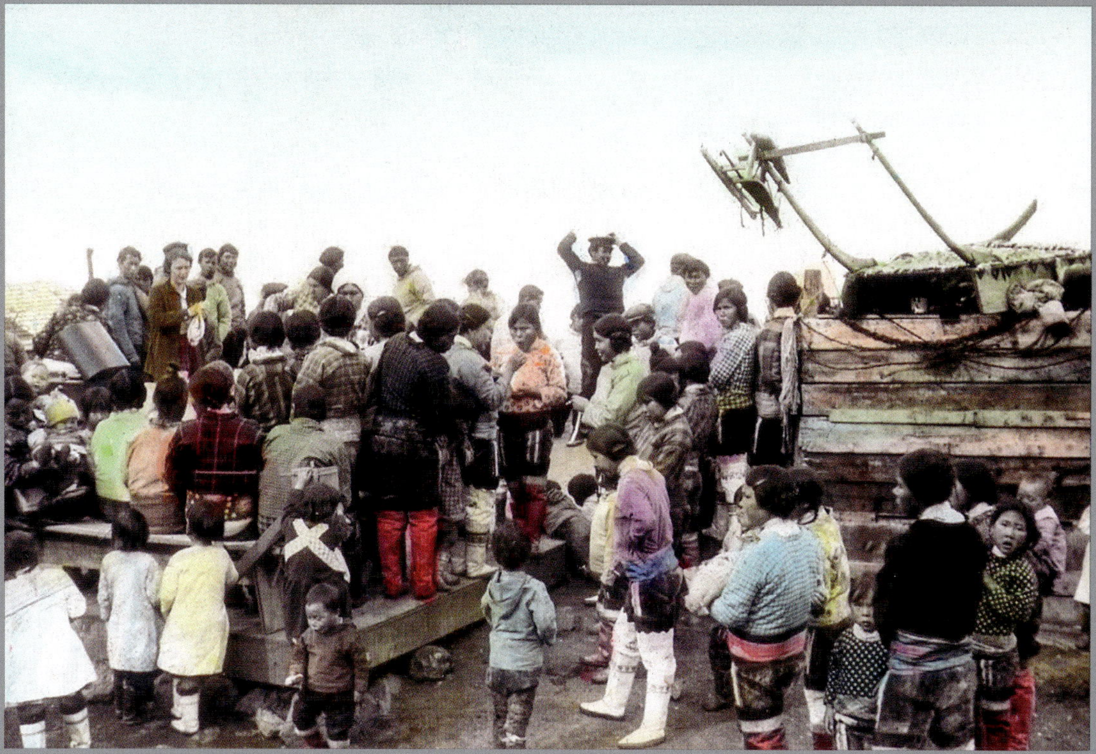

*ca.* 1932

The Danish ship *Disko* leaving Sisimiut Harbor. *RK-6096, Plattsburgh State Art Museum*

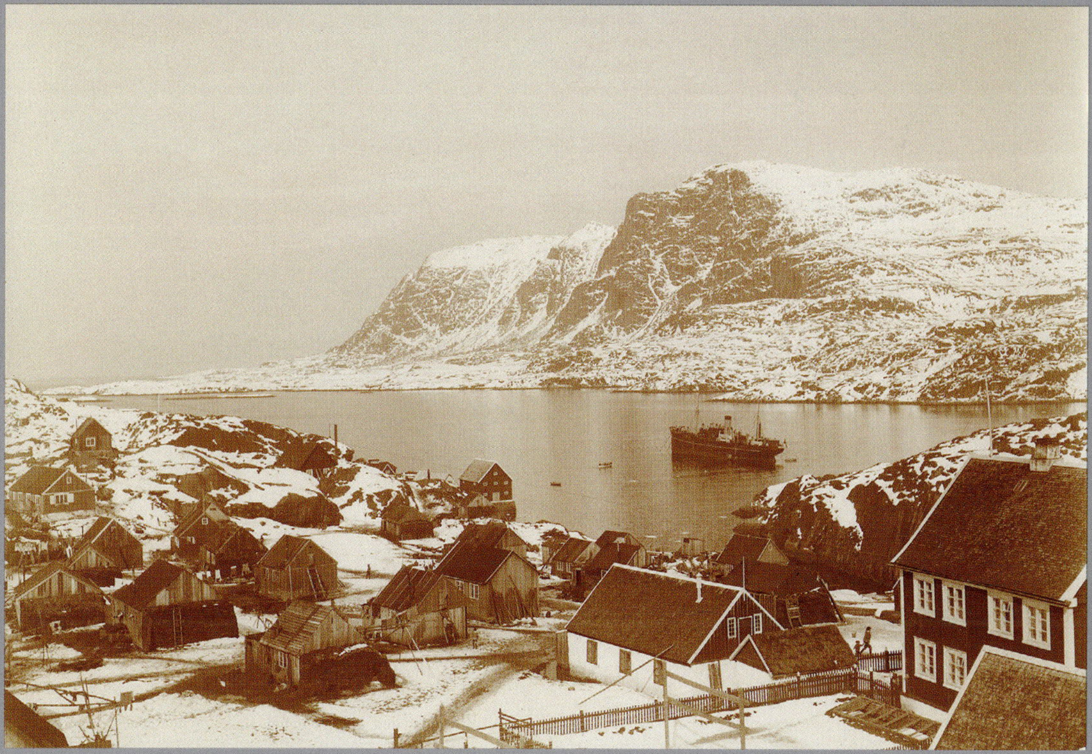

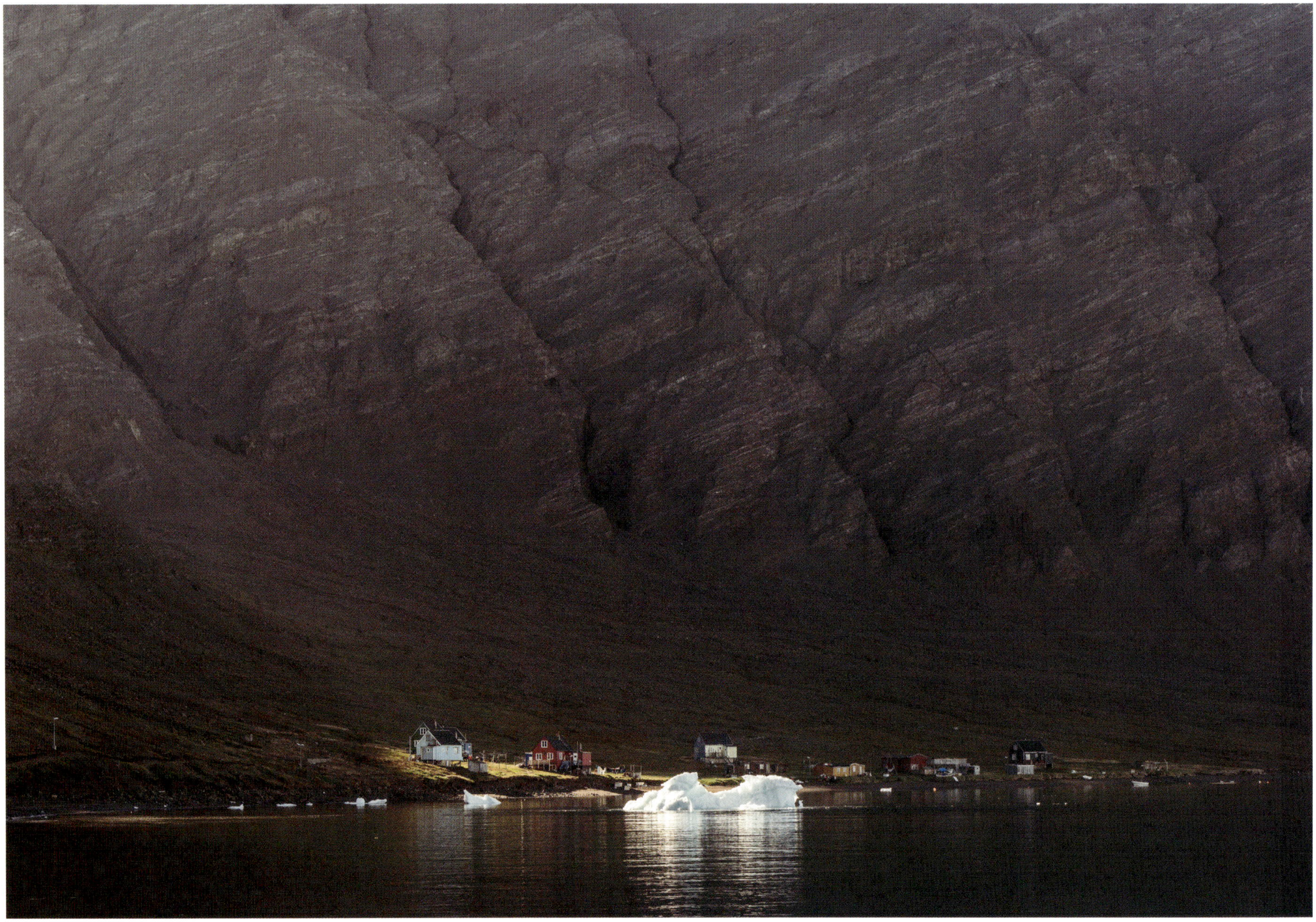

ILLORSUIT 2017

The last light of the evening sun illuminates the strip of land with four or five houses composing the "northern suburbs" of Illorsuit.

# Kent and Defibaugh in Illorsuit, Greenland

AXEL JEREMIASSEN

**Greenland, and Illorsuit particularly, connects two American photographers separated by nearly a century. The photographers share a magnetic fascination with the island and its people, as expressed through their pictures.**

Though today perhaps better known for his paintings and graphic design, Rockwell Kent made numerous photographic images of Greenland in the 1930s; contemporary photographer Denis Defibaugh, in residence from 2016 to 2017, made many more. Often, Kent used his photographs as references later put into service for his paintings. Defibaugh's images serve artistic, documentary, and anthropological purposes, as much for the present as the future. Viewed together, as the present book affords, the images make understandable the photographers' fascination. Kent's and Defibaugh's work offers today's viewers insights to—and some understanding of—Greenlandic moments as seen, captured, and selected by the makers and their medium. As always, viewers do not know what happened before and after the shutter snapped shut, never mind what might have gone on outside the frame. Though precisely exact and unblinking, images alone cannot hope to tell Greenland's robust story.

Illorsuit (spelled *Igdlorssuit* in 1930) is a Greenlandic word meaning big houses. Perhaps intentionally ambiguous, the reference may be either to the traditional Greenlandic houses made of sod and stones, or to buildings made by the Royal Trade Department (KGH). Geographically, the settlement is bounded by a large fjord complex along the west-central Greenland coast, populated by many glaciers and islands. Hunters were attracted by the tremendous number of sea animals in the area both during the summer and winter seasons. Illorsuit officially became a trading station in 1859, more than 10 years after the hunters took up residence on the island. In the early 1930s, Kent's time, hunters sought seals, sharks, belugas, and narwhals as a source of income for their families. Male family members were groomed from a young age to become hunters who could supply their families and relatives with game. Women processed seal skins and sewed clothes, tents, *qajaq* (kayaks), and other necessities; they prepared whale and seal fat as oil for lamps and fuel to heat houses. Commercially, a hunter's catch could be sold to the KGH.

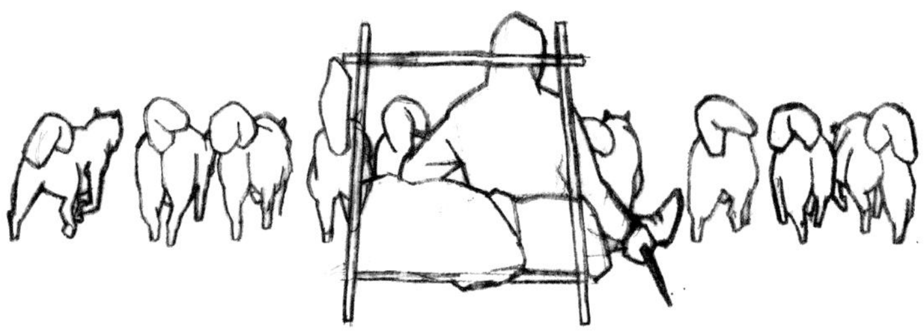

In the 1930s, the KGH was the only place where settlers could sell the products of their hunting, and it was the only place where they could buy necessities such as rifles, cartridges, knives, and cotton, and indulgences such as European coffee, candies, and tobacco.

The phenomenon of the Arctic's "Midnight Sun" begins in mid-May, lasting through July. As the long summer days bring warm weather to Illorsuit, it is time to relax. People sleep less, camp, travel the fjords by boat, and hunt and fish at all hours of the day. High mountains that hide the sun mean the polar darkness begins in November, lasting until the beginning of February. Winter is the indoor time of year, while hunting and fishing are still daily rituals.

Countless generations have adapted over the years to living in Greenland's extreme environmental circumstances. The coast and fjords of North Greenland from the Sisimiut area are covered by ice all winter, from as early as November or December right up until April, May, or even June, depending on weather, geographical conditions, and sea currents.

Mother Nature's unpredictability required hunters to use sledge dog teams instead of kayaks during these frozen months. In the 1930s, hunters used at least eight dogs to pull their sledges on the sea ice. Each family depended on seal hunting not only for the skin, but for the year-round supply of food and fuel they provided. Ringed seal were the most common sea animal, but other types of seals were hunted as well, along with eiders, guillemots, seagulls, and black-legged kittiwakes, the latter coveted birds. Fishing provided income and food for families: Capelin, wolffish, cod fish, halibut, red fish, Greenland halibut, trout/char, salmon, and others were plentiful in Greenland's icy waters.

In the summer, hunters turned to their kayaks for hunting and fishing, and Inuit boys learned how to handle these boats from a very young age. During winter, they traveled to the edge of the ice by dog sledge to then hunt from their kayaks.

Over time, the number of seals in the area declined as did the demand for their hides. The once-prestigious occupation of Illorsuit hunter with kayak and sledge dogs faded away and, by the early 1960s, was abandoned in favor of halibut fishing with the long line as the substitute income producer. Initially, halibut fishers used rowing boats; today, swift, open motor boats are favored. Fish factories built in the Uummannaq district, roughly 60 miles southeast, became buyers for the yield. While there are still many kayaks in use, today they are used primarily for recreation.

Governing managers administered the larger settlements where early 20th-century Greenland inhabitants lived. Whether an oupost or trading station, the red KGH buildings symbolized the colonial administration establishment. Beside the manager, typically a Lutheran protestant minister, a catechist, a Danish female teacher to instruct the colonial administration officials' children, a baker, a carpenter and cooper who built wooden barrels for the shipping of fat, a nurse or midwife, and occasionally a doctor also lived in and supported the community. Larger-sized settlements typically included such structures as a KGH store, church, school, port, and a warehouse. Smaller outposts featured a small store and warehouse; a school and church shared one building. The smallest settlements, trading stations, were inhabited by fewer than 100 locals living in houses often made of stone and soil, and rarely had trading or KGH official facilities.

Officially, Greenland's Danish colony status ended in 1953; unofficially, it continued until the introduction of home-rule government in 1979. Then, Greenland introduced its own legislative assembly, establishing a government elected by Greenlanders. The government made Greenlandic the public national language (2009) and helped create and develop museums, a national theater, radio/TV broadcasting, and an educational system that includes a social service training college and the University of Greenland. While the government helped build national pride, still, politicians remain challenged by the nation's expansive landscape and the many small, isolated communities. The availability and distribution of goods, modernizing transportation systems and health care, and increasing the number of qualified educators remain goals for Greenland's leadership.

A massive landslide fell to the sea in the barren Karrat Fjord on Greenland's west coast on June 17, 2017. The landslide caused a sudden surge of seawater—a "mega-tsunami," one of the tallest in recorded history—that wreaked havoc in the fishing settlement of Nuugaatsiaq, located on an island within the fjord about 12 miles away from the landslide. The 300-foot-high wave washed away 11 houses, devastating Nuugaatsiaq. Although the tsunami dissipated quickly as it crossed the deep, four-mile-wide fjord, it still had enough energy to send water more than 50 yards up the Illorsuit hillside on the opposite side of the fjord. Almost all of the fishermen lost their boats and all equipment. Houses, buildings and other public facilities were damaged and, in some cases, destroyed or taken away by the powerful waves. Due to the danger of additional landslides from the unstable mountain, the government closed both settlements and moved the populations to Uummannaq, Ilulissat, and Aasiaat. Most Illorsuit residents moved to Uummannaq. It is unknown if Illorsuit and Nuugaatsiaq ever will be populated again. Seismologists who studied the event warn that such events could become more frequent as the climate warms. Hermann Fritz, an environmental engineer at the Georgia Institute of Technology, led the tsunami research team. "With global warming and sea level rise," he said, "such landslides are going to be far more common."[1]

Today, some might say photographers Kent and Defibaugh "parachuted in," tourists no matter any other intentions. Their work thematically overlaps sometimes, other times diverging sharply, each mirroring much in the same way that Greenland's broad history evolved. The island nation's people and their work and social habits, political and community organizations, unsurprisingly morph over time. Remaining ever constant is Greenland's extraordinary, often overwhelmingly stark, and sometimes unforgiving environment, shaping and at least occasionally being shaped by—or maybe being adapted to by—its inhabitants and the Greenlanders' responses to modernization.

**Note**

1 "Huge Landslide Triggered Rare Greenland Mega-Tsunami," *Nature: International Weekly Journal of Science*, (July 31, 2017): http://www.nature.com/news/huge-landslide-triggered-rare-greenland-mega-tsunami-1.22374.

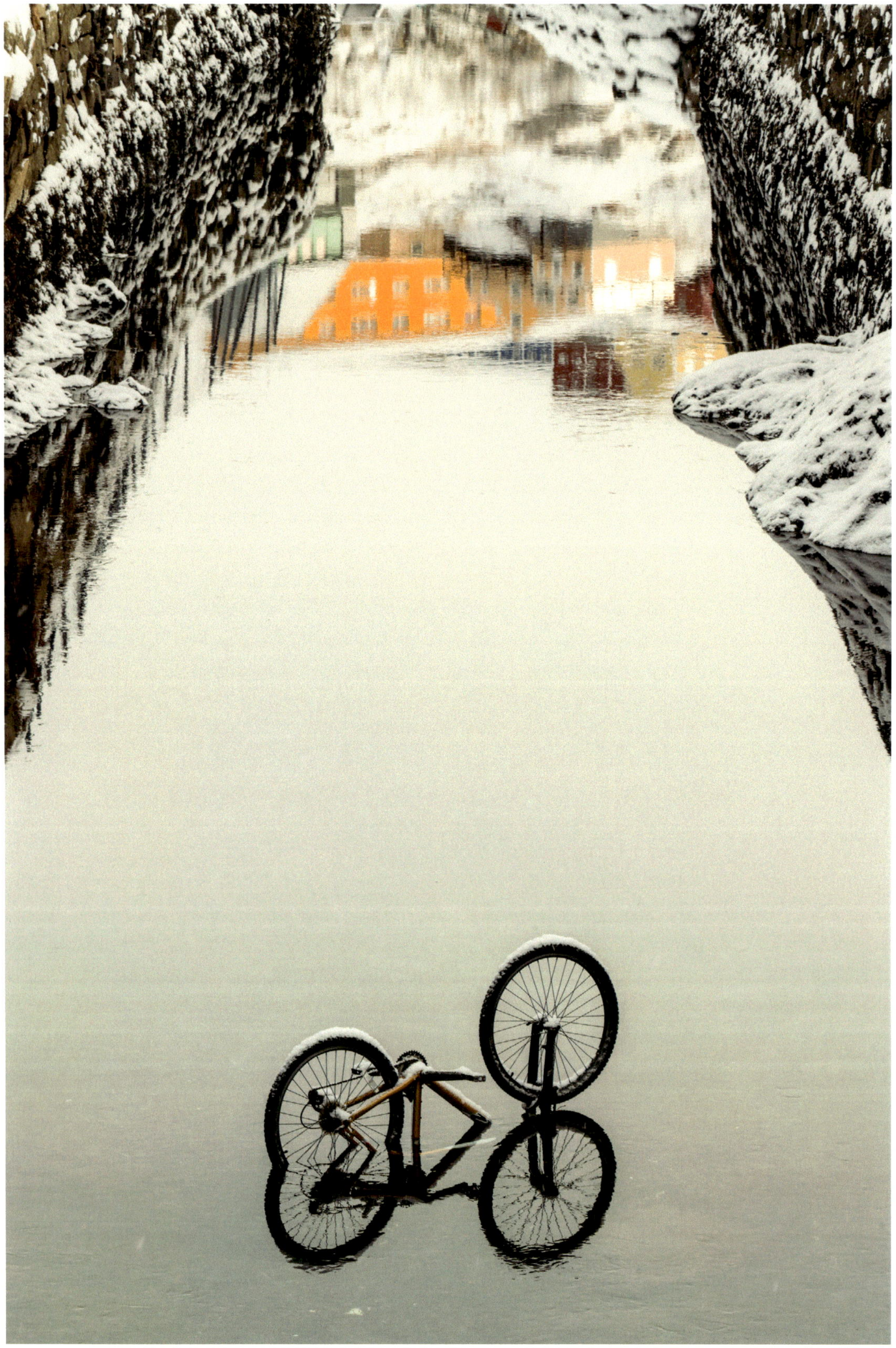

SISIMIUT 2016
This short, nameless canal in Sisimiut often becomes the final resting spot for bikes.

# About an American who Spent a Year in Illorsuit 1931–32[1]

SAMUEL MØLLER

*Translated to English by Axel Jeremiassen*

**During the 1930s and Danish colonial rule, Greenland was a "closed" country: non-Danes could not enter without governmental permission. Greenlanders would not have known any foreigners except Danish public employees and those rare explorers who may have traveled along the coast in the late 19th and early-20th centuries.**

Some "Letters to the Editor" in the two national periodicals, *Avangnâmioĸ* and *Atuagagdliutit*, described foreign fishing ships belonging to Norwegians, French, and Portugese, but these ships were never allowed to port at any settlements, except to obtain water or in case of emergencies.

Catechist Samuel Møller's (1885–1961) account about Rockwell Kent for *Avangnâmioĸ*, translated to English and presented below, is the only article published by either of the periodicals, although one might have expected at least some account or note about the 1929 wreck of an American ship south of Nuuk. As Møller's article reports, the locals knew of and were fond of Rockwell Kent and were particularly excited by his visit, since it was extremely rare that foreign ship accidents occurred at that time; nor was it typical for foreigners to live by themselves in such isolated communities. The article implies that Kent had contacted the appropriate Danish authorities to request permission to stay by himself in Greenland.

*Axel Jeremiassen*

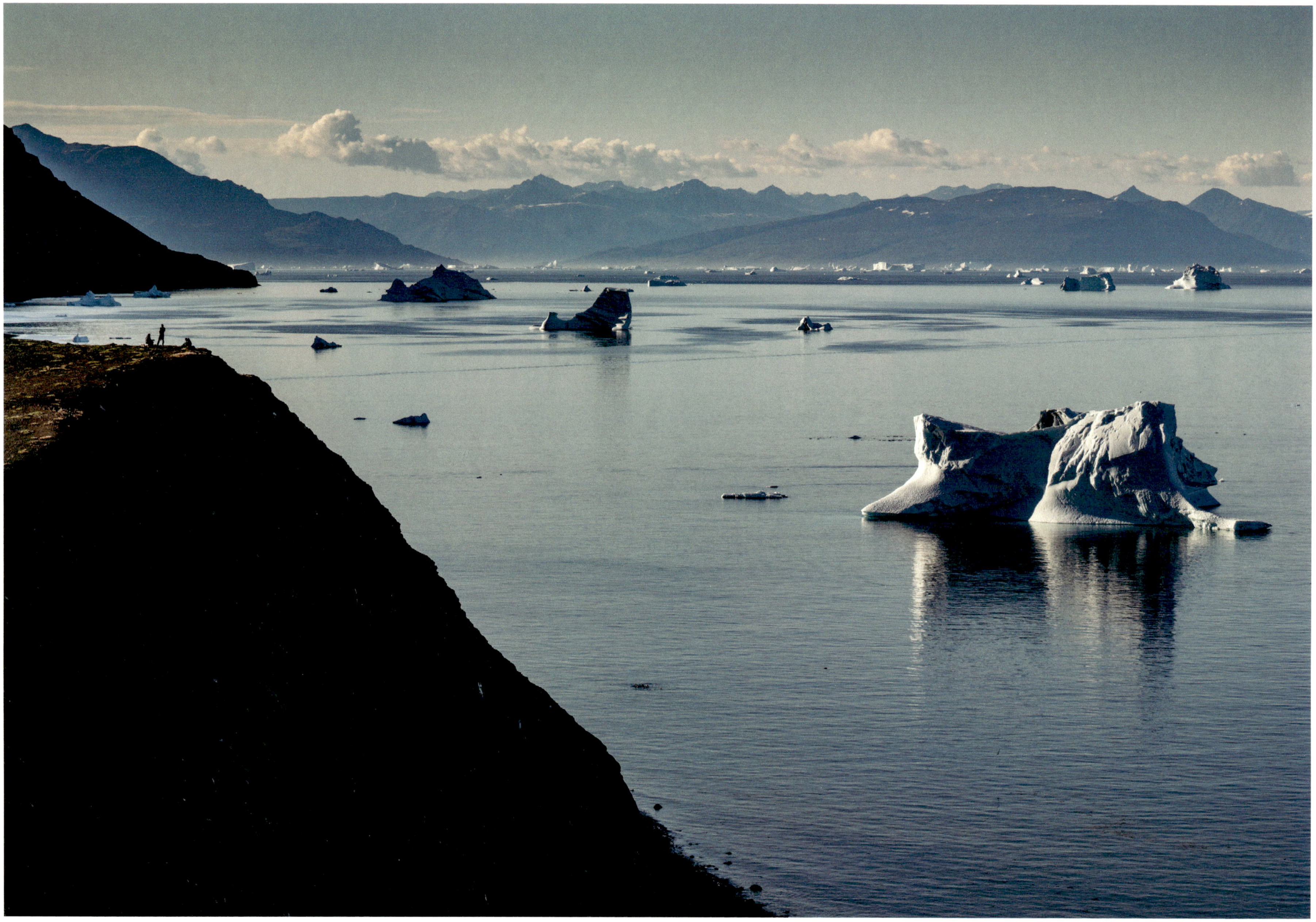

ILLORSUIT 2017

The Illorsuit Sound and the headlands of Illorsuit. A perfect spot to view the Sound and surrounding mountains and glaciers.

In early 1931, while on my way to a North Greenland Provincial Council Assembly, I arrived in Uummannaq one day before the sailing ship, *Disko*. I had heard about an American sailing on that ship who was going to spend a year in Illorsuit; but before I saw him, we left for Qeqertarsuaq through Upernavik.

Upon my return, I met the man I had heard about. He was very friendly; but since I did not know his language, I could not speak with him, even though he seemed kindly. I learned about a house being built with a plinth made of cement, and that had been placed higher on the hillside than other houses. I realized that Rockwell Kent had started construction by himself with a few helpers. Since he was very skilled at carpentry, he soon finished his house—and it was very beautiful. Kent soon became like one of our own citizens, even though he was quite different. We did not feel shy around him and he really did not act like someone who was from another country.

When we played ball, we took him down many times—although he was strong, he was not well prepared. We saw his artistic skills through his paintings, and he really *could* paint; I wonder how many times he painted the other side of our settlement? During sunsets, when the sun was no longer in sight, he painted the red-colored mountains beautifully. He often painted very big scenes, and infrequently, small ones; during the fall, he continued his painting while also taking photos.

When he finished building his house, he invited me to visit. I saw many books on his desk and stacks of completed papers that he had written. I realized that he was not only an artist, but an author as well. He might have discovered many things to write about while he was here, and for sure, it will be good for us—the way Greenlanders live is not "superficial." But there probably will be some things he will write about us that will not be so good as well!

In late summer, he went to Uummannaq a few times, since he had a rented motorboat that he used to go to a place north of us called Umiivik to paint landscapes. He had intended to travel by sledge dog teams to see more landscapes, although he barely traveled during winter, except to the Nuugaatsiaq area to paint the landscapes. I have seen several of his paintings and can say that, as an artist, he is someone whose talents cannot be ignored.

Kent always accepted invitations and was always included by those celebrating birthdays, since he always gave excellent gifts. I'm wondering what he thought about Greenlandic coffee, although he never said no to a coffee. He often invited those he wanted to see to visit him at his house; and most of the time, people accepted his invitations, since he never was reluctant to share his good taste in food and drink.

He also tried to share happiness—no wonder he was so wealthy. A widow from Uummannaq was his housekeeper while he was here. When he had his birthday (June 21), we visited first; while we had a coffee, he came in with a big cake with his name on it. When he cut it up, he told us we would find something inside—one guest found 50 cents, another 25 cents, another 5 cents. My wife found 2 cents; I did not find anything, even though I tried. Those who went later on expressed the great fun when a hag found some money and started screaming in a dialect that did not sound like Illorsuit's. I never counted the amount of coins, but there were many more when his wife had her birthday.

When Christmas was near, he announced that he would give everyone a meal and he started his preparations. He gave food to all of the children first, and then he invited elders to his house to feast. He gave us a meal, alcoholic beverages, and gifts.

When someone asked to borrow money from him, he never said no—in fact, he let hunters borrow a lot of money during the bad season. In addition, when you did minor work for him, he paid very well. I once got 10 kr. for making inner windows for his house. His sailboat pilot got 100 kr. His helpers had a lot to thank him for—his generosity allowed them to buy things they could not have afforded if not for him. While he was here, a few people died; he gave 10 kr. to those who lost someone.

At the beginning of 1932, he went to Nuugaatsiaq, accompanied by a hunter who was going to teach him how to run sledge dog teams. When he returned from that visit, we learned how people in Nuugaatsiaq were doing. Afterward, he began going by himself to closer areas; and during the spring, he sometimes went with someone else.

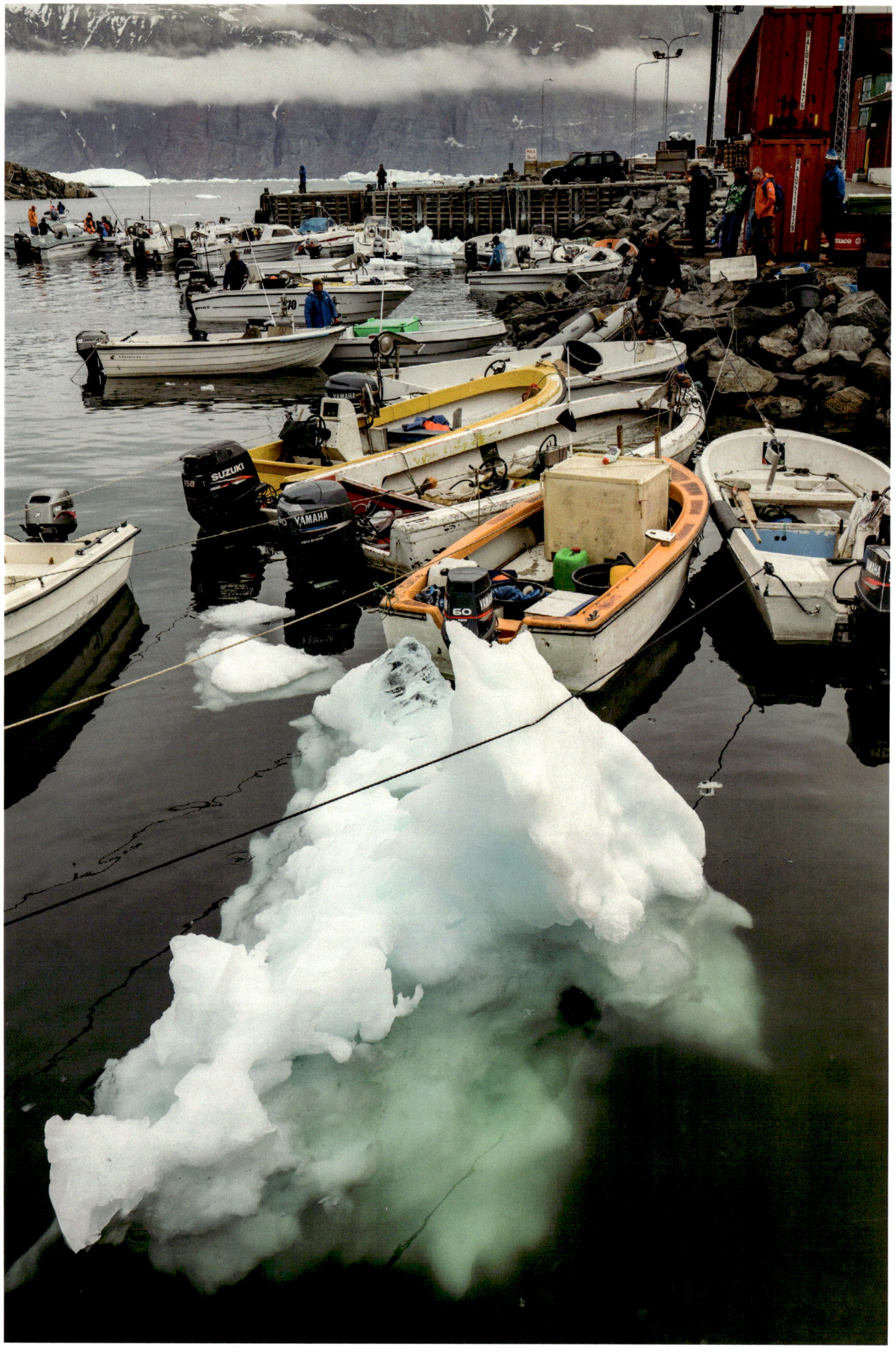

UUMMANNAQ 2017
Uummannaq Harbor bustling with fishermen and boats.

One time he announced that his wife was coming; and he went to Qeqertarsuaq to pick her up by sledge dog. At the time, you had to go through Kangerlussuaq, which is a long way away; since it was thawed, the closer path is not covered by ice. He did not even reach Uummannaq, but instead had to go through the East to finish his trip by boat to reach Qeqertarsuaq.

While he was out of the settlement, his beloved ones missed him, since he had promised to come back. He did come back to our settlement by motorboat in May. His wife, whom we saw for the first time, greeted everyone politely. After she arrived, Kent announced that he was going to build a dancing house, and before his wife's birthday, he made an outdoor floor. Nevertheless, when there was open sea, he started sailing to Nuussuaq and even wanted to reach Kullorsuaq, a settlement much farther north. In September, they finished the outdoor floor with help from hunters and others. Of course, since there were several carpenters, they finished it in no time. What a wonderful house it was—about 6 meters forward and 5 meters to the side. The outside color was light blue and the top was coated by tarpaper. Its doors, which did not reach the ground, were openable during the summer, and on the upper side was a propeller, a gift from a German pilot named Ernst Uder. There were big windows on both sides of his house, east and west, and Kent carved a piece of wood that says:

*Æ*
*1932*
*Igdlorsuarne igdlo*
*Atautsimìtarfik ĸitigtarfiuvdlunilo*

A loose translation of this inscription is, *"Established 1932. A house in Illorsuit, meeting and dancing house."*

When Kent's house was finished in late September, the locals wanted to thank Kent and his wife and so they invited them for a coffee. Three Germans who had been living nearby were invited as well. It was a shame that we had not prepared properly, since Kent had announced that he and his wife were leaving in two days. We wanted to use the thank-you visit for their goodbye party. At the time, because we had a large assembly of people and the house was packed, I asked our municipality representatives to offer some remarks, which they agreed to do. I decided to also offer some remarks myself, even though I did not feel ready to do so. This is what I said:

*"This evening, we honor Kent and his wife, along with these Germans whom we have invited to thank for a gift. However, we have not been able to prepare properly in such a short time. It is a fact that those who go forward—compared to those of us from the North—may never again see such a fine house...Moreover, this room is so beautiful...and we did not offer any payment, and offered nothing but 'thank you.' We will keep Kent and his wife in our memories and we thank them for their lavish gift to us. As long as this room exists, it will remind us of Kent and his wife. Also, inhabitants of the other settlement who visit and see this room—but who will never meet Kent—will understand Kent's kindness, generosity, and carpentry skills in building this dance hall for the community. We also thank those Germans who are here for their donation as well. In addition, we thank Kent for his many gifts and for all the festivities he hosted while he was here; and the money he donated to relatives when a family member died. While they journey to their homeland, we wish them both safety and a blessed trip. We who live here will not forget the Kents, for their gift lies in the center of our settlement."*

After my speech, we sang their favorite hymn. And then a young fellow hunter—on behalf of the youth—made a short speech.

Kent thanked everyone for their words and we expressed our thankfulness for such a big room that was so needed. Later, I learned that [Danish explorer and anthropologist] Knud Rasmussen had donated some equipment for the house.

The following day, the Kents left around 11 a.m. by Uummannaq motorboat, and were saluted with rifles and hymns. I think the best way to describe Rockwell Kent and what he meant to the people of Illorsuit is to say that he was kindly to all and that because of that, he often was called, "The man who was respected by the locals and became part of our community."

**Note**

1 Samuel Møller's original article was published in *Avangnâmioĸ* as "Amerikamiup Rockwell Kent-ip Igdlorssuarne ukînera uk. 1932," N. 12, December 1932.

# Acknowledgments

Special thanks to those from the Greenland communities of Illorsuit, Uummannaq, Sisimiut, and Nuuk as well as all the organizations and individuals that provided support and advice for this project. For providing institutional foundation support, I thank the National Science Foundation's Arctic Social Sciences Program, Rochester Institute of Technology, Ilisimatusarfik (the University of Greenland), and SUNY Binghamton.

Anna Kerttula de Echave, former program director for the Arctic Social Sciences Program, supported and encouraged this project from its beginning. Colleen Strawhacker, Ph.D., current program director, Arctic System Sciences and Arctic Social Sciences, Arctic Sciences Section, Office of Polar Programs National Science Foundation, also continues to provide support. Axel Jeremiassen provided a wealth of knowledge about Greenlandic culture and did an unimaginable amount of translating and interpreting. Christine Løventoft, former Sisimiut Museum director, enthusiastically supported the project's community engagement.

For their willingness to take part in this adventure, I thank my partners: Axel Jeremiassen, Susan Vanek, and Jette Rygaard. Appreciation as well to students Kyra Krenitsky and Alexandra Pasquale.

Charlie Bush of RIT/NSF Sponsored Research kept the project proposal moving forward and it was refined through Rebecca Sumner's editing. Michele Washburn continually kept me organized; and Robin Amico kept my budget in line.

Thank you to Cecilia Esposito, Karen Blough, Felicia Herzog, and Rebecca LeClair for introducing the Rockwell Kent lantern slides to me. Your helpfulness and education about the fabulous Rockwell Kent Collection at Plattsburgh State Art Museum was invaluable.

I acknowledge the kind sponsorship of Kristine Bosworth of Nikon Inc. for the donation of equipment and for the shock-resistant cameras for the Inuit student photo workshop; Samantha Corn and Sony Electronics Inc. for video support; and Canon for portable printers. Kyli Cosper and Polar Field Services, Inc. provided logistical support for lodging and transport in Greenland.

Invaluable insight and advice during the photo editing process was provided by Whitney Tressel, Gregory Halpern, Ken Geiger, Jasna Bogdanovska, and John Aäsp.

My friends and colleagues at RIT's School of Photographic Arts and Sciences were supportive of the creative process in more ways than they can imagine. Thank you, Therese Mulligan, Doug Manchee, Rachel Ferraro, Clay Patrick McBride, Patti Russotti, Susan Lakin, Roberley Bell, David Walter, Bill DuBois, Dawn Tower DuBois, and William Snyder.

Much respect and deep appreciation to the students, teachers, and good people of Greenland for their kindness and support of the project. In particular, Ann Andreasen and the folks at the Uummannaq Children's Home were always welcoming and invited me to new adventures. While there were many I met who shared their Greenland experiences and knowledge, I acknowledge in particular Loogki Fleischer, Niels Peter Geisler, Riikki Gronvold, Kulunnguaq Jeremiassen, Benjamin Jonaghansen, Jarte Korneliussen, Kasper Korneliussen, Johan Fredrek Kreutzmann, Nuka-Sofie Fleischer Lovstrom, Hans Mørch, Mathias Nielsen, Olafur Rafnar Olafsson, Dorthe Qvist, Erik Qvist, Agathe and Jan Storch, Ester Zeeb, Katrina Zeeb, Pavia Zeeb, and Knud Zeebi.

Thanks to Bruce Austin, director of RIT Press, for his enthusiasm, attention to detail, and tenacious effort to complete the book properly and on time. And to designer Heidi Trost, copy editor Kathleen S. Smith, and production manager Marnie Soom for their contributions to the book.

Additional valued contributions to the project and book were provided by Jamie Allen, Erika Larsen, Mark Nuttall, Gretel Ehrlich, Rick Wester, Wendy Marks, and Elaine Defibaugh through their experience, expertise, creativity, and dedication to art.

# Colophon

*Editor* Bruce A. Austin

*Design* Heidi Trost

*Production* Marnie Soom

*Typefaces* Whitney, Surveyor

*Paper* McCoy Silk 100-pound text

*Printing and binding* Jostens Book Manufacturing

Funding for this book made possible by a grant from the National Science Foundation. Additional funding support generously provided by the Rochester Institute of Technology's School of Photographic Arts and Sciences and by friends of RIT Press: Jon Budington, John Williams, and Kathy Yu.

This first edition of *North by Nuuk, Greenland after Rockwell Kent* is limited to 1250 casebound copies.